Collins

AQA GCSE
Science (9–1)
Maths in Science Practice Pack

Amanda Clegg and Karen Collins

William Collins' dream of knowledge for all began with the publication of his first book in 1819. A self-educated mill worker, he not only enriched millions of lives, but also founded a flourishing publishing house. Today, staying true to this spirit, Collins books are packed with inspiration, innovation and practical expertise. They place you at the centre of a world of possibility and give you exactly what you need to explore it.

Published by Collins
An imprint of HarperCollins*Publishers*
The News Building, 1 London Bridge Street, London, SE1 9GF, UK

HarperCollins*Publishers*
Macken House, 39/40 Mayor Street Upper, Dublin 1, D01 C9W8, Ireland

Browse the complete Collins catalogue at
collins.co.uk

10 9 8 7 6 5 4 3 2 1

A catalogue record for this publication is available from the British Library.

ISBN 978-0-00-878318-1

Authors: Amanda Clegg and Karen Collins
Publisher: Katie Sergeant
Product manager: Jessica Ashdale
Development editor: Sarah Ryan
Copyeditor: Mike Harman
Proofreader: Aidan Gill
Cover designer: Gordon MacGilp
Cover image: Susana Guzman / Alamy
Internal designer and illustrator: Six Red Marbles, India
Typesetter: Mike Harman
Production controller: Alhady Ali
Printed and bound in the UK by Ashford Colour Ltd

collins.co.uk/sustainability

Acknowledgements

The publishers gratefully acknowledge the permission granted to reproduce the copyright material in this book. Every effort has been made to trace copyright holders and to obtain their permission for the use of copyright material. The publishers will gladly receive any information enabling them to rectify any error or omission at the first opportunity.

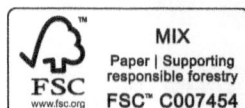

Contents

HT: Higher tier only

Introduction

Examiners' reports repeatedly show that students struggle to answer maths in science questions at both GCSE and A level. In addition, research indicates that students find it difficult to transfer knowledge they acquire in mathematics to their science lessons, even when they perform well in maths and physics (Woolnough, 2000). We have designed this book to support you in helping students transfer and apply their mathematical knowledge to a science context.

This book is split into 5 chapters:

1. Arithmetic
2. Handling data
3. Algebra
4. Graphs
5. Geometry and trigonometry

Each chapter consists of worked examples showing how to carry out the mathematical skill. This is followed by at least one faded example. Fading is recommended by the Educational Endowment Fund (EEF) as a form of scaffolding to support students once they have experienced worked examples. The steps are removed in reverse order (fading from the bottom) to provide greater support for novice learners (Pritchard, 2022). Finally, there are student questions, using the appropriate mathematical skill, written in a biology, chemistry and physics context, where applicable. Some mathematical concepts are only assessed in one or two of the sciences, and this has been reflected in the text. The student questions increase in level of difficulty to provide differentiated practice.

How to use this book

An important consideration is diagnosis. Which skills are students finding difficult? How do you know? These activities provide useful exercises for both intervention and revision of key mathematical skills in context. They also show the links between the three sciences as well as between mathematics and science.

Once an area of concern has been identified you could use the exercises in this text to:

- Model how to carry out the mathematical skill using the worked example, 'thinking aloud' to explain the metacognitive process you are going through whilst completing the activity. Clear modelling of the thinking, and the questions we ask ourselves when solving a problem, demonstrates to students how an expert would approach it (Mulholland, 2022).
- Work through the faded example(s) together, modelling the initial steps and asking students to complete the faded sections. Guide and scaffold the students' practice.
- Students complete the subject-specific questions independently to consolidate their learning. Provide support where needed.

This approach starts with direct instruction, using small steps to avoid overloading the students' working memory, and finishes with the students carrying out independent deliberate practice.

All the introductory material, questions and answers, and PowerPoint slides of the worked and faded examples can be downloaded in editable format at www.collins.co.uk/MathsInScience/download.

Woolnough, J. (2000). *'How do students learn to apply their mathematical knowledge to interpret graphs in physics?'*, Research in Science Education, 30, (3), 259–267.

Mulholland, K. (2022). *'Thinking Aloud to support mathematical problem-solving.'* [1st February 2025] online. Available from: https://educationendowmentfoundation.org.uk/news/eef-blog-thinking-aloud-to-support-mathematical-problem-solving

Pritchard, B. (2022). *'Ways into Science – Making the most of worked examples.'* [11th January 2025] online. Available from: https://educationendowmentfoundation.org.uk/news/eef-blog-ways-into-science-making-the-most-of-worked-examples

Chapter 1 Arithmetic

Standard form

Scientists use standard form to represent very large or very small numbers. Standard form is written in the form of $a \times 10^n$ where a is a number between 1 and 10 ($1 \leq a < 10$), multiplied by a power of ten.

When multiplying standard form: multiply the numbers (a) and add the powers (n). When dividing standard form: divide the numbers (a) and subtract the powers (n).

Worked example

Q1. A comet travels at a speed of 2×10^4 m/s.

Calculate the time it takes the comet to travel 2×10^6 metres. [3 marks]

Write down the equation:	$\text{time} = \dfrac{\text{distance}}{\text{speed}}$	
Substitute variables into equation:	$\dfrac{2 \times 10^6}{2 \times 10^4}$	[1 mark]
When dividing numbers in standard form we divide the numbers and subtract the powers:	$1 \times 10^{6-4}$	[1 mark]
Write the answer:	1×10^2 s	[1 mark]

Faded examples

Q1. A wave travels at a speed of 3×10^8 m/s in air. The frequency of the wave is 3×10^9 Hz.

Calculate the wavelength in air. [3 marks]

Write down the equation:	$\text{wavelength} = \dfrac{\text{wave speed}}{\text{wave frequency}}$	
Substitute variables into equation:	$\dfrac{3 \times 10^8}{3 \times 10^9}$	[1 mark]
Divide the numbers and subtract the powers:	... ...	[1 mark]
Write the answer:	...	[1 mark]

Q2. The length of a cell in an image is 2×10^{-4} m. The real length of the cell is 2×10^{-5} m.

Calculate the magnification. [3 marks]

Write down the equation:	$\text{magnification} = \dfrac{\text{size of image}}{\text{size of real object}}$	
Substitute variables into equation:	...	[1 mark]
Divide the numbers and subtract the powers: ...		[1 mark]
Write the answer:	...	[1 mark]

Biology questions

Q1. There are approximately 5 000 000 red blood cells in a drop of human blood.

Write the number 5 000 000 in standard form. .... [1 mark]

Q2. There are approximately 86 000 000 000 neurones in the human brain.

Write the number 86 000 000 000 in standard form. .... [1 mark]

Q3. A chloroplast has a diameter of 0.000005 m.

Write the number 0.000005 in standard form. .... [1 mark]

Q4. A red blood cell has a diameter of 0.00000085 m.

Write the number 0.00000085 in standard form. .... [1 mark]

Q5. A person in the UK needing a blood transfusion has a probability of 5×10^{-7} of contracting Hepatitis C.

Write this probability as a decimal. .... [1 mark]

For Question 6 use the following equation: $\quad \text{magnification} = \dfrac{\text{size of image}}{\text{size of real object}}$

Q6. The length of a cell in an image is 4×10^{-4} m. The real length of the cell is 2×10^{-4} m.

Calculate the magnification. ....

....

....[3 marks]

Q7. A typical human body contains approximately 3.0×10^{13} cells. Each human body cell contains approximately 2.0×10^4 protein molecules.

How many protein molecules are in a typical human body approximately? ....

....

....[3 marks]

Q8. A typical leaf contains 2.0×10^5 chloroplasts. In optimal conditions, each chloroplast contributes to the production of approximately 1.0×10^2 glucose molecules per hour.

Calculate how many glucose molecules could be produced per leaf. ....

....

....[3 marks]

Q9. Cell A has length 4.2×10^{-5} m and cell B has length 1.4×10^{-6} m.

Calculate how many times larger cell A is compared to cell B. ....

....

....

....[3 marks]

[Total marks 17]

Chemistry questions

Q1. Buckminsterfullerene has the formula C_{60}. The relative formula mass of buckminsterfullerene is 720.

Write the number 720 in standard form. .... [1 mark]

Q2. A substance has a mass of 40 000 g.

Write the number 40 000 in standard form. .... [1 mark]

Q3. A solution has a concentration of 0.001 g/dm^3.

Write the number 0.001 in standard form. .... [1 mark]

Q4. 0.025 dm^3 of liquid was transferred to a conical flask using a pipette.

Write the number 0.025 in standard form. .... [1 mark]

Q5. The radius of a sodium ion is approximately 0.000000098 m.

Write the number 0.000000098 in standard form. .... [1 mark]

For Questions 6 and 7 use the following equation: $\text{concentration} = \dfrac{\text{mass of solute}}{\text{volume of solution}}$

Q6. A solution contains 2.5×10^{-2} g of solute in 5.0×10^{-2} dm^3 of water.

Calculate the concentration of the solution in g/dm^3. ....

....

....[3 marks]

Q7. A student dissolved 0.1 g of a substance in 2.5×10^{-2} dm^3 of water.

Calculate the concentration of the solution in g/dm^3. ....

....

....[3 marks]

For Questions 8 and 9 use the following equation:

mass of substance = amount of substance (moles) × relative formula mass

Q8. Sodium has a relative formula mass of 23. Calculate the mass of 1×10^{-3} moles of sodium.

Give your answer in standard form. ....

....

....   g   [3 marks]

Q9. Boron has a relative formula mass of 11. Calculate the mass of 5.5×10^{-3} moles of boron.

Give your answer in standard form. ....

....

....   g   [3 marks]

[Total marks 17]

Physics questions

Q1. The frequency of FM radio waves is 100 000 000 Hz.

Write the number 100 000 000 in standard form. .... [1 mark]

Q2. The pressure in the deepest part of the ocean is approximately 110 000 000 Pa.

Write the number 110 000 000 in standard form. .... [1 mark]

Q3. The wavelength of some visible rays is 0.0000005 m.

Write the number 0.0000005 in standard form. .... [1 mark]

Q4. The radius of a nucleus of an atom is approximately 0.0000000001 m.

Write the number 0.0000000001 in standard form. .... [1 mark]

For Questions 5, 6 and 7 use the following equation: $\text{wavelength} = \dfrac{\text{wave speed}}{\text{wave frequency}}$

Q5. Light has a wave speed of 3×10^8 m/s in air. The frequency of green light is 5.5×10^{14} Hz

Calculate the wavelength of the green light in air ....

....

....[3 marks]

Q6. The frequency of some electromagnetic waves is 2.4×10^9.
The waves have a speed of 3×10^8 m/s in air.

Calculate the wavelength. ....

....

....[3 marks]

Q7. An electromagnetic wave has a wave speed of 3×10^8 m/s in air.
The frequency of the wave is 3×10^{-7} Hz.

Calculate the wavelength. ....

....

....[3 marks]

[Total marks 13]

Ratios

A ratio is a way of comparing two quantities. It shows how much there is of one item compared to another. Ratios are written using a colon, such as 3 : 1. It's important that the numbers are written in the correct order. For example, if you are asked for the surface area to volume ratio then the numbers must be written in this order: surface area : volume. Units are not needed as a ratio is a comparison of two similar quantities. The simplest ratio is when the numbers have no common factors other than 1 and cannot be further simplified.

Worked examples

Q1. A bag contains 12 red beads and 3 yellow beads.
What is the simplest whole number ratio of yellow to red beads? [1 mark]

Write down the ratio in words: yellow : red

Substitute values into ratio: 3 : 12

Simplify: 1 : 4 [1 mark]

Q2. In mice, black fur (B) is dominant over white fur (b). Two heterozygous mice (Bb) with black fur are crossed. The possible genotypes are shown below in the Punnett square.
What is the phenotypic ratio of black fur to white fur offspring? [1 mark]

	B	b
B	BB	Bb
b	Bb	bb

Write down the possible genotypes of the offspring: BB   Bb   Bb   bb

Write down the phenotype of each offspring (visible trait): black black black white

Calculate the phenotypic ratio and write in the same order as the question: 3 : 1 [1 mark]

Faded examples

Q1. A bag contains 15 red beads, 10 orange beads and 5 blue beads.
What is the simplest whole number ratio of red to orange beads? [1 mark]

Write down the ratio in words: red : orange

Substitute values into ratio: 15 : 10

Simplify: ... .. [1 mark]

Q2. A bag contains 15 red beads, 10 orange beads and 5 blue beads.
What is the simplest whole number ratio of blue to red beads? [1 mark]

Write down the ratio in words: blue : red

Substitute values into ratio: ... ..

Simplify: ... .. [1 mark]

Ratios

Q3. In mice, black fur (B) is dominant over white fur (b). One heterozygous mouse (Bb) with black fur and a homozygous dominant (BB) are crossed. The possible genotypes are shown below in the Punnett square.

What is the phenotypic ratio of black fur to white fur offspring? [1 mark]

	B	B
B	BB	BB
b	Bb	Bb

Write down the possible genotypes of the offspring: BB BB Bb Bb

Write down the phenotype of each offspring (visible trait): black black black black

Calculate the phenotypic ratio and write in the same order ... [1 mark]
as the question:

Q4. In mice, black fur (B) is dominant over white fur (b). One heterozygous mouse (Bb) with black fur and a homozygous recessive are crossed. The possible genotypes are shown below in the Punnett square.

What is the phenotypic ratio of black fur to white fur offspring? [1 mark]

	b	b
B	Bb	Bb
b	bb	bb

Write down the possible genotypes of the offspring: Bb Bb bb bb

Write down the phenotype of each offspring (visible trait): ...

Calculate the phenotypic ratio and write in the same order
as the question: ... [1 mark]

Biology questions

Q1. A mouse has a heart rate of 600 beats per minute and a human heart beats 75 times per minute.

What is the simplified ratio of the mouse's heart rate to the human's?..

.... [1 mark]

Q2. The table below shows measurements for three cube-shaped unicellular organisms.
Calculate the surface area to volume ratio for each organism. [3 marks]

Unicellular organisms	Surface area (mm^2)	Volume (mm^3)	Surface area to volume ratio (mm^2/mm^3)
A	6	1	
B	24	8	
C	96	64	

Q3. In pea plants the allele for tallness (T) is dominant over the allele for shortness (t). A heterozygous plant (Tt) is crossed with a homozygous recessive plant (tt). Here is a Punnett square showing the possible offspring genotypes.

What is the phenotypic ratio of tall to short plants? .... [1 mark]

	T	t
t	Tt	tt
t	Tt	tt

Q4. A pond water sample was placed on a microscope slide. A student counted 12 green algal cells and 6 protozoa (one-celled organisms) in the field of view.

What is the ratio of algae cells to protozoa?... [1 mark]

Q5. Look at the balanced symbol equation for photosynthesis: $6CO_2 + 6H_2O \rightarrow C_6H_{12}O_6 + 6O_2$

What is the ratio of molecules of carbon dioxide used to glucose produced? .... [1 mark]

Q6. Polydactyly is an inherited condition where a person has an extra finger or toe. It is caused by inheriting a single dominant allele (A). Someone who is homozygous recessive (aa) will not have polydactyly.
Use the genetic diagrams to write down the phenotypic ratio of having polydactyly to not having it.

a)

	A	a
a	Aa	aa
a	Aa	aa

b)

	A	A
a	Aa	Aa
a	Aa	Aa

c)

	A	a
A	AA	Aa
a	Aa	aa

....

....[3 marks]

[Total marks 11]

Chemistry questions

Q1. The formula for carbon dioxide is CO_2.

What is the ratio of carbon to oxygen in a molecule of carbon dioxide?....................................... [1 mark]

Q2. The formula for ammonium nitrate is NH_4NO_3.

What is the ratio of nitrogen to oxygen in the formula for ammonium nitrate?............................. [1 mark]

Q3. Copper oxide reacts with hydrogen to produce copper and water: $CuO + H_2 \rightarrow Cu + H_2O$

What is the ratio of hydrogen gas to water in the reaction?.. [1 mark]

Q4. Sulfuric acid reacts with sodium hydroxide in a neutralisation reaction.

$H_2SO_4 + 2NaOH \rightarrow Na_2SO_4 + 2H_2O$

What is the ratio of sulfuric acid to sodium hydroxide in the reaction? .... [1 mark]

Q5. A purple dye contains 65% red dye and 35% blue dye.

What is the simplest whole number ratio of red dye to blue dye in the purple dye? .... [1 mark]

Q6. What is the simplest ratio of carbon to hydrogen atoms in the following hydrocarbons?
    a) C_2H_4 **b)** C_6H_{14} **c)** $C_{16}H_{34}$ **d)** $C_{45}H_{90}$ [1 mark each]

    a) ......   **c)**   ...

    b) ...   **d)**   ...

Q7. The surface area of a cube is 24 nm^2. The volume of the cube is 8 nm^3.

What is the simplest whole number surface area to volume ratio of the cube? .... [1 mark]

Q8. The surface area of a cube is 1.5 cm^2. The volume of the cube is 0.125 cm^3.

What is the simplest whole number surface area to volume ratio of the cube? .... [1 mark]

Q9. Determine the empirical formula of the following ionic compound.   [1 mark]

    ◯ Cl^-

    ● Na^+

[Total marks 12]

Physics questions

Q1. In the nucleus of an atom there are 86 protons and 136 neutrons.

What is the simplest whole number proton to neutron ratio in the nucleus?.................................. [1 mark]

Q2. A gear with 36 teeth drives a gear with 12 teeth. The gear ratio is the relationship between the number of teeth on two interlocking gears.

What is the simplest whole number gear ratio for these gears?.. [1 mark]

Q3. A transformer has 1200 turns in the primary coil and 400 turns in the secondary coil.

What is the simplest whole number turns ratio of primary turns : secondary turns?...................... [1 mark]

Q4. Here is a table showing the energy transferred per second from a house.

Roof	Windows	Floor	Walls	Door
1.8 kJ/s	1.2 kJ/s	0.8 kJ/s	1.2 kJ/s	1.0 kJ/s

What proportion of the energy was transferred through the walls? .... [1 mark]

[Give your answer as the simplest whole number ratio of energy transferred through the walls to total energy transfer.]

Q5. A motor has an efficiency of 85%.
Calculate the simplest whole number ratio of useful energy output to wasted energy output.

.... [1 mark]

Q6. Two springs have spring constants of 300 N/m and 50 N/m.
If they are extended by the same amount, what is the simplest whole number ratio of elastic potential energy stored in the stiffer spring compared to the less stiff spring?

.... [1 mark]

Q7. The percentage of electricity generated by gas-fired power stations in 2018 was 32%.
This dropped to 30% in 2024.
What is the simplest whole number ratio of electricity generated by gas-fired power stations in 2018 compared to 2024?

.... [1 mark]

Q8. A star has 25 times more mass compared to the mass of the Sun.
What is the ratio of the mass of this star compared to the Sun? .... [1 mark]

Q9. Jupiter has 63 known moons and Uranus has 27.
What is the simplest whole number ratio of known moons from Jupiter compared to Uranus?

.... [1 mark]

Q10. A series circuit contains three resistors with resistances of 12 Ω, 18 Ω and 6 Ω. The potential difference across each resistor is proportional to its resistance.
If the total potential difference across the circuit is 24 V, calculate the simplest whole number ratio of potential differences across the three resistors.

....

....[2 marks]

[Total marks 11]

Percentages and percentage change

A percentage is a fraction that is written as a number of parts per hundred. One 'per cent' means that you have one part per hundred. There are many potential situations where you may be asked to calculate a percentage or percentage change.

When calculating percentage change you should indicate if there has been an increase or a decrease using a positive or negative sign. So +12.4% shows an increase and –5.3% shows a decrease.

Worked examples

Q1. Find 25% of 300 N. [1 mark]

Divide 25 by 100 to give a decimal: $\frac{25}{100} = 0.25$

Multiply the decimal by 300 N: $0.25 \times 300 = 75$ N [1 mark]

Q2. **Chemistry only** A reaction has a theoretical yield of 10.4 g of product. The actual yield was 8.2 g. Calculate the percentage yield. [2 marks]

Write down the equation:
$$\text{percentage yield} = \frac{\text{mass of product actually made}}{\text{maximum theoretical mass of product}} \times 100$$

Substitute values into the equation: $= \frac{8.2 \text{ g}}{10.4 \text{ g}} \times 100$ [1 mark]

Calculate the answer: $= 78.8\%$ [1 mark]

Q3. A potato disc measured 5.0 cm in diameter. It was placed in distilled water for 30 minutes. The diameter was re-measured at 5.3 cm. Calculate the percentage change. [2 marks]

Calculate the difference in diameter: $5.3 - 5.0 = 0.3$ cm

Write down the equation:
$$\text{percentage change} = \frac{\text{difference}}{\text{original}} \times 100$$

Substitute values into the equation: $= \frac{0.3}{5.0} \times 100$ [1 mark]

Calculate the answer and indicate if it is an increase (+) or a decrease (–): $= +6\%$ [1 mark]

Q4. **Chemistry only** Calculate the atom economy for the production of ethanol (C_2H_5OH) from this chemical reaction:

$$C_2H_4 + H_2O \rightarrow C_2H_5OH \qquad A_r\text{: H = 1; C = 12; O = 16}$$ [3 marks]

Write down the equation:
$$\text{atom economy} = \frac{\text{relative formula mass of desired product}}{\text{sum of relative formula masses of all reactants}} \times 100$$

Calculate M_r of product and all reactants:
$C_2H_5OH = (2 \times 12) + (6 \times 1) + 16 = 46$ [1 mark]
$C_2H_4 + H_2O = (2 \times 12) + (6 \times 1) + 16 = 46$

Substitute values into the equation: $= \frac{46}{46} \times 100$

Calculate the answer: $= 100\%$ [1 mark]

Faded examples

Q1. A student investigated the effect of solution concentration on the mass of uncooked potato pieces. The mass at the start was 5.2 g and the mass after 20 minutes was 5.5 g. Calculate the percentage increase in mass. **[2 marks]**

Calculate the difference: $\qquad$ 5.5 g – 5.2 g = 0.3 g

Divide the difference by the original: $\qquad \dfrac{0.3}{5.2} = 0.06$ **[1 mark]**

Multiply by 100 to find the percentage change: ... **[1 mark]**

Q2. An investigation was carried out to investigate osmosis in carrots. The length of a piece of carrot at the start was 2.0 cm. After 20 minutes the length increased to 2.1 cm. Calculate the percentage increase in length. **[2 marks]**

Calculate the difference: $\qquad$ 2.1 cm – 2.0 cm = 0.1 cm

Divide the difference by the original: ... **[1 mark]**

Multiply by 100 to find the percentage change: ... **[1 mark]**

Biology questions

Q1. There were 120 tadpoles in a small pond. After seven days, 30 of them had grown legs and become frogs. What percentage of the tadpoles had turned into frogs?

.... [1 mark]

Q2. There were 20 tomato plants in a greenhouse. 15 of them were flowering. What percentage of tomato plants have flowers?

.... [1 mark]

Q3. Look at this food chain. The energy (J) present in each trophic level of the food chain is displayed below:

lettuce → slug → small bird → large bird of prey

Energy (J)   50 000   5000   500   50

Calculate the percentage of energy originally present in the lettuce plants which was transferred to the bird of prey.

.... [1 mark]

Q4. A student places a potato chip weighing 1.2 g into a boiling tube containing salt solution. After 30 minutes the student removes the chip and blots it dry. The chip is reweighed and has a final mass of 1.02 g. Calculate the percentage change in mass for the potato chip.

....

....[2 marks]

Q5. A cylinder of carrot weighs 0.85 g. It is placed in a boiling tube containing distilled water. After 30 minutes the carrot cylinder is removed and blotted dry. It now has a mass of 0.97 g.
Calculate the percentage change in mass of the carrot cylinder.

....

....[2 marks]

Q6. A bacterial culture initially contains 250 000 cells. After treating with an antibiotic for 24 hours, 175 000 cells survive.
What percentage of the bacterial population was killed by the antibiotic?

....

....[2 marks]

Q7. A sample of pond water contained 400 photosynthetic microorganisms. It was exposed to sunlight for 3 hours and the number increased to 550 microorganisms.
What was the percentage increase in the microorganism population?

....

....[2 marks]

[Total marks 11]

Chemistry questions

Questions 1, 2 and 3 use the equation for percentage mass, which is written below for Q1:

$$\text{percentage mass} = \frac{\text{mass of carbon atoms in CO}}{\text{mass of all atoms in CO}} \times 100$$

Q1. The formula for carbon monoxide is CO. The relative atomic masses (A_r) for the elements in carbon monoxide are: O = 16, C = 12.

Calculate the percentage by mass of carbon in carbon monoxide. ....

....[2 marks]

Q2. The formula for ammonia is NH_3. The relative atomic masses (A_r) for the elements in ammonia are: N = 14, H = 1.

Calculate the percentage by mass of nitrogen in NH_3. ....

....[2 marks]

Q3. The formula for ammonium nitrate is NH_4NO_3. The relative atomic masses (A_r) for the elements in ammonia are: O = 16, N = 14, H = 1.

Calculate the percentage by mass of nitrogen in NH_4NO_3. ....

....[2 marks]

For Question 4 use the following equation:

$$\text{percentage yield} = \frac{\text{mass of product actually made}}{\text{maximum theoretical mass of product}} \times 100$$

Q4. The theoretical yield is 1500 g.
Calculate the percentage yield for each mass of product made.   [2 marks each]

a) 500 g   **b)** 750 g   **c)** 1000 g   **d)** 1200 g

a)   ....   c)   ....

b)   ....   d)   ....

For Questions 5 and 6 use the following equation:

$$\text{atom economy} = \frac{\text{relative formula mass of desired product from equation}}{\text{sum of relative formula masses of all reactants from equation}} \times 100$$

Q5. The relative formula mass of the desired product of a reaction is 112. The relative formula mass of all the reactants was 244.
Calculate the percentage atom economy. ....

....[2 marks]

Q6. Sodium carbonate can be made using the Solvay process: $2NaCl + CaCO_3 \rightarrow Na_2CO_3 + CaCl_2$
Calculate the atom economy for the production of sodium carbonate. The relative formula masses for the compounds are: Na_2CO_3 = 106, NaCl = 58.5, $CaCO_3$ = 100

....

....

....[3 marks]

[Total marks 19]

Physics questions

Q1. A light bulb converts 15% of the electrical energy it receives into light energy.

What percentage of energy is wasted? .... [1 mark]

Q2. In our solar system there are eight planets and four of these planets are gas giants.

What percentage of the planets in our solar system are gas giants? .... [1 mark]

Q3. Terawatt-hours (TWh) are used to measure energy production and consumption. Renewable energy sources have generated a record level of the UK's electricity recently.

Year	Wind (TWh)	Solar (TWh)	Nuclear (TWh)
2024	84.1	14.8	40.6
2023	82.3	13.9	40.6

Calculate the increase as a percentage from 2023 to 2024 for:

a) wind   **b)** solar   **c)** nuclear   **d)** total of all three sources.   [2 marks each]

a) ....   **c)** ....

b) ....   **d)** ....

Q4. A transformer steps down voltage from 240 V to 12 V.

Calculate the percentage reduction in voltage..

....[2 marks]

Q5. An electric motor has an efficiency of 85%. If 200 J of electrical energy is supplied to the motor, how much useful mechanical energy is produced?   [2 marks]

For Questions 6, 7 and 8 use the following equation:

$$\text{percentage efficiency} = \frac{\text{useful energy output}}{\text{total energy input}} \times 100$$

Q6. The total energy input for a lightbulb is 200 J. The useful energy output of the bulb is 70 J.
Calculate the percentage efficiency of the lightbulb. ....

....[2 marks]

Q7. The output for a set of speakers is 500 J. The total energy input is 850 J.
Calculate the percentage efficiency of the speakers. ....

....[2 marks]

Q8. A solar panel receives 1200 watts of solar energy and converts it into 264 watts of electrical energy. The manufacturer claims the solar panel has an efficiency of 25%.
Find the difference between the actual efficiency and the manufacturer's figure.

....

....

....[2 marks]

[Total marks 20]

Rates

The rate is the speed at which something happens or changes or the number of times it happens or changes in a period of time. Generally, the following equation can be used to calculate rate:

$$\text{rate of change} = \frac{\text{change in value}}{\text{change in time}}$$

There are no questions in this section for physics.

Worked examples

Q1. A student ran on a treadmill for 10 minutes. The student's heart beats 270 times in 3 minutes. Calculate the student's average heart rate in beats per minute. [2 marks]

Write down the equation:	$\text{average heart rate} = \dfrac{\text{change in value}}{\text{change in time}}$	
Substitute values into equation:	$\text{average heart rate} = \dfrac{270 \text{ beats}}{3 \text{ minutes}}$	[1 mark]
Calculate answer and give units:	$\dfrac{270}{3} = 90$ beats per minute	[1 mark]

Q2. A student investigated the reaction between zinc and hydrochloric acid. After 50 seconds, 40 cm^3 of hydrogen gas was produced. This increased to 55 cm^3 after 100 seconds. Calculate the average rate of reaction between 50 s and 100 s. [2 marks]

Write down the equation:	$\text{average rate} = \dfrac{\text{change in value}}{\text{change in time}}$	
Substitute values into equation:	$\text{average rate} = \dfrac{55 - 40 \text{ cm}^3}{100 - 50 \text{ s}}$	[1 mark]
Calculate answer and give units:	$\dfrac{15}{50} = 0.3$ cm³/s	[1 mark]

Faded examples

Q1. A student used an exercise machine. The student's heart rate was monitored. The heart beat 280 times in 2 minutes. Calculate the student's average heart rate in beats per minute. [2 marks]

Write down the equation:	$\text{average heart rate} = \dfrac{\text{change in value}}{\text{change in time}}$	
Substitute values into equation:	$\text{average heart rate} = \dfrac{280 \text{ beats}}{2 \text{ minutes}}$	[1 mark]
Calculate answer and give units:	… ..	[1 mark]

Q2. A student used an exercise machine. The student's breathing rate was monitored. The student took 22 breaths in 2 minutes. Calculate the student's average breathing rate in breaths per minute. [2 marks]

Write down the equation:	$\text{average rate} = \dfrac{\text{change in value}}{\text{change in time}}$	
Substitute values into equation:	… ..	[1 mark]
Calculate answer and give units:	… ..	[1 mark]

Rates

Q3. A student investigated the rate of reaction between marble chips and acid. They measured the mass of the reaction over time. Here are the results of their investigation.

Time (s)	Mass of flask and contents (g)
0	152.0
20	150.4
40	147.8
60	144.4

Calculate the average rate of reaction between 20 seconds and 60 seconds. [2 marks]

Write down the equation:
$$\text{average rate} = \frac{\text{change in value}}{\text{change in time}}$$

Substitute values into equation:
$$\text{average rate} = \frac{150.4 - 144.4 \text{ g}}{60 - 20 \text{ s}}$$
[1 mark]

Calculate answer and give units: … .. [1 mark]

Q4. A student investigated the reaction between hydrochloric acid and calcium carbonate. They measured the volume of gas produced over time. Here are the results of their investigation.

Time (s)	Volume of hydrogen (cm³)
0	0
20	26
40	40
60	48
80	50

Calculate the average rate of reaction between 40 seconds and 80 seconds. [2 marks]

Write down equation:
$$\text{average rate} = \frac{\text{change in value}}{\text{change in time}}$$

Substitute values into equation: … .. [1 mark]

Calculate answer and give units: … .. [1 mark]

Biology questions

Q1. A student measured their heart rate over 7 minutes. They counted 525 beats.

Calculate the students heart rate in beats per minute. ....

....[2 marks]

For Question 2 use the equation: transpiration rate = $\dfrac{\text{volume of water used}}{\text{time}}$

Q2. A student investigated transpiration by measuring the uptake of water by a plant. In 30 minutes, the plant took up 6 cm³ of water.
Calculate the rate of water uptake in cm³ per minute. ....

....[2 marks]

Q3. A student ran for 5 minutes and during this time a heart rate monitor recorded 650 beats. After resting for 10 minutes, the monitor had recorded 720 beats.

a) Calculate the student's heart rate during exercise in beats per minute. ....

....[2 marks]

b) Calculate the student's heart rate during the rest period in beats per minute. ....

....[2 marks]

Q4. A student investigated the rate of an enzyme reaction by measuring the volume of gas produced. At the start no gas had been produced. After 2 minutes, 8 cm³ of gas had been collected. After 5 minutes, a total of 15 cm³ of gas had been collected.

a) Calculate the rate of reaction at 2 minutes. ....

....[2 marks]

b) Calculate the rate of reaction at 5 minutes. ....

....[2 marks]

Q5. A student investigated the rate of photosynthesis in a water plant by counting the number of oxygen bubbles released in different light intensities over 5 minutes. The light intensity was changed by moving a lamp different distances from the plant.

Distance from lamp (cm)	Number of oxygen bubbles in 5 minutes
0	0
10	115
20	55

a) Calculate the rate of oxygen bubble production when the lamp was 10 cm away................................

....[2 marks]

b) Calculate the rate of oxygen bubble production at a distance of 20 cm. ....

....[2 marks]

Rates

Q6. Students used a potometer to measure the rate of water uptake by a woody stem. They recorded the distance a water bubble moved along a scale. The bubble moved a distance of 68 mm in 9 minutes.

What was the rate of water uptake per minute for this stem?..

....[2 marks]

Q7. A doctor was investigating blood flow through two blood vessels A and B. In blood vessel A, it took 10 seconds for 15 cm^3 of blood to travel through. In blood vessel B, it took 15 seconds for 12 cm^3 of blood to pass through.
Calculate the rate of blood flow through both blood vessels and ensure the units are correct.

....

....

....[4 marks]

[Total marks 22]

Chemistry questions

Q1. During a reaction between marble chips and hydrochloric acid, 16 cm³ of hydrogen gas was produced in 20 seconds.

Calculate the rate of reaction...

....[2 marks]

Q2. During a reaction between zinc and sulfuric acid, 70 cm³ of hydrogen gas was produced in 100 seconds.

Calculate the rate of reaction...

....[2 marks]

Q3. During a reaction between calcium carbonate and hydrochloric acid, 0.008 mol of hydrogen gas was produced in 16 seconds.

Calculate the rate of reaction...

....[2 marks]

Q4. A student investigated the rate of reaction between marble chips and hydrochloric acid. The mass at the start of the investigation was 152 g. This dropped to 151 g after 200 seconds.

Calculate the average rate of reaction...

....[2 marks]

Q5. Here are the results from an investigation into the reaction between marble chips and hydrochloric acid.

Time (s)	Volume of gas (cm³)
20	16
100	48

a) Calculate the rate of reaction at 20 seconds. ....

....[2 marks]

b) Calculate the average rate of reaction between 20 seconds and 100 seconds. ....

....[2 marks]

Q6. A student measured the mass of copper produced during the electrolysis of copper sulfate.

Time (min)	Mass of copper (g)
10	0.12
20	0.24

a) Calculate the rate of reaction after 10 minutes..

....[2 marks]

b) Calculate the average rate of reaction between 10 minutes and 20 minutes...................................

....[2 marks]

[Total marks 16]

Chapter 2 Handling data

Significant figures

When carrying out a measurement, the number of significant figures indicates how precise the measurement is. When we carry out a calculation, the number of significant figures normally reflects the precision of the values used in the calculation.

When rounding a value to a number of significant figures we need to look at the 'decider digit'. If the decider digit is 5 or above then you would round up, if below 5 then you would round down. Care needs to be taken here. Many marks are lost on examination papers because students make errors when rounding.

Worked examples

Q1. Write the number 356 700 to three significant figures. [1 mark]

Identify the 'decider digit':	As we are rounding to three significant figures the decider digit is the fourth digit: 356 **7**00
Decide if we need to round up or round down:	The decider digit is more than 5 so we need to round up the 6 to a 7.
Calculate answer (include units if given):	357 000 [1 mark]

Q2. Write the number 0.02403 m to three significant figures. [1 mark]

Identify the 'decider digit':	0.024**0**3 (the first significant figure is the 2)
Decide if we need to round up or round down:	Round down as the decider digit is less than 5.
Calculate answer (include units if given):	0.0240 m [1 mark]

Faded examples

Q1. Write the number 0.935 A to two significant figures. [1 mark]

Identify the 'decider digit':	0.93**5** (the first significant figure is the 9)
Decide if we need to round up or round down:	Round up as the decider digit is 5.
Calculate answer (include units if given):	... [1 mark]

Q2. Write the number 753 876 m to two significant figures. [1 mark]

Identify the 'decider digit':	75**3** 876
Decide if we need to round up or round down:	...
Calculate answer (include units if given):	... [1 mark]

Biology questions

Q1. Complete the table and write each number to the required number of significant figures.

[1 mark each row]

	Number	To three significant figures	To two significant figures
a)	9762		
b)	3725		
c)	0.2657		
d)	3.705		
e)	0.003725		

Q2. A student calculated that there were 1678.2 daisy plants in a field.

Give this answer to three significant figures. .... [1 mark]

Q3. The mean number of stomata on a leaf was calculated as 45.75.

Give this answer to two significant figures. .... [1 mark]

Q4. A student calculated the area of a piece of land to be 21 876 m². Give this answer to:

a) three significant figures   **b)** two significant figures   [1 mark each]

....

Q5. The mean reaction time for a student in an investigation was 0.1627 s. Give this answer to:

a) three significant figures   **b)** two significant figures   [1 mark each]

....

Q6. A student calculated the mean body temperature of the class as 37.42 °C.

Give this answer to three significant figures. .... [1 mark]

Q7. A blood sample contains 186 432 red blood cells. Give this value to:

a) three significant figures   **b)** two significant figures   [1 mark each]

....

Q8. There were 2.847×10^3 bacteria in a sample solution.

Give this answer to three significant figures. .... [1 mark]

[Total marks 15]

Chemistry questions

Q1. Complete the table and write each number to the required number of significant figures.

[1 mark per row]

	Number	To three significant figures	To two significant figures
a)	5876		
b)	4735		
c)	0.3457		
d)	6.507		
e)	0.002325		

Q2. A student calculated the percentage by mass of potassium in potassium hydroxide as 69.642857%.

Give this answer to two significant figures.. [1 mark]

Q3. The concentration of a solution of sodium chloride solution was calculated to be 58.44 g/dm^3.

Give this answer to three significant figures. .... [1 mark]

Q4. The average mass of copper produced during the electrolysis of copper sulfate was 3.325 g.

Give this answer to:

a) three significant figures   **b)** two significant figures   [1 mark each]

....

Q5. The maximum mass of product that could be produced in a reaction was 1.66667 g. Give this answer to:

a) three significant figures   **b)** two significant figures   [1 mark each]

....

Q6. The concentration of a solution was calculated to be 0.26271 mol/dm^3.

Give this answer to three significant figures. .... [1 mark]

Q7. The mean volume of gas produced in 60 seconds was 0.046043 dm^3. Give this value to:

a) three significant figures   **b)** two significant figures   [1 mark each]

....

Q8. A neutron has a mass of 1.6749×10^{-27} kg.

Give this value to two significant figures.. [1 mark]

[Total marks 15]

Physics questions

Q1. Complete the table and write each number to the required number of significant figures.

[1 mark each]

	Number	To three significant figures	To two significant figures
a)	7354		
b)	5825		
c)	0.4637		
d)	6.305		
e)	0.007625		

Q2. A student calculated a change in gravitational potential energy as 18.380508 J.

Give this answer to three significant figures. .... [1 mark]

Q3. The acceleration of an object was calculated to be 1.66667 m/s^2.

Give this answer to three significant figures. .... [1 mark]

Q4. The resistance of a heating element was calculated to be 8.2142857 Ω. Give this answer to:

a) three significant figures   **b)** two significant figures   [1 mark each]

....

Q5. The mean time taken for an activity was calculated to be 8.8565 s. Give this answer to:

a) three significant figures   **b)** two significant figures   [1 mark each]

....

Q6. The power output of a bulb was calculated to be 130.5 W.

Give this answer to three significant figures. .... [1 mark]

Q7. A student found the density of water was 0.9982 g/cm^3 at 20 °C. Give this value to:

a) three significant figures   **b)** two significant figures   [1 mark each]

....

Q8. The half-life of potassium-40 is 1.4010×10^{10} years.

Give this value to three significant figures. .... [1 mark]

[Total marks 15]

Arithmetic mean

The arithmetic mean is often simply called the mean. It is the average of several values found by adding them together and then dividing by the number of values. Odd results or anomalies can skew the mean, so we identify anomalies and investigate to see if they were caused by errors or are genuine data. Remove errors before calculating a mean, but keep all the data in the results table and highlight them as anomalies.

Worked examples

To calculate the mean of a set of data, add up all the individual values and then divide by the total number of values.

$$mean = \frac{sum\ of\ values}{number\ of\ values}$$

Q1. A student counts the number of petals on five different daisy flowers.
The results were: 13, 15, 14, 12 and 16.
Calculate the mean number of petals per daisy flower. [2 marks]

Calculate the sum of the individual petal counts: $13 + 15 + 14 + 12 + 16 = 70$ [1 mark]

Divide this by the number of values to find the mean: $\frac{70}{5} = 14$ [1 mark]

Q2. A student carried out a titration. The table shows the student's results.

	Trial 1	Trial 2	Trial 3	Trial 4
Volume of solution (cm^3)	39.20	38.25	38.35	38.30

Only use the concordant data (results within 0.10 cm^3 of each other). [3 marks]

Identify the concordant data: 38.25 cm^3, 38.35 cm^3, 38.30 cm^3 [1 mark]

Find the sum of the concordant data: $38.25 + 38.35 + 38.30 = 114.90$ [1 mark]

Divide this by the number of values to find the mean: $\frac{114.90}{3} = 38.30\ cm^3$ [1 mark]

Faded examples

Q1. A plant scientist is studying the effect of a new fertiliser on the growth of tomato plants. The height increase over 4 weeks was measured with the standard fertiliser and the new fertiliser.

Plant number	Height increase with standard fertiliser (cm)	Height increase with new fertiliser (cm)
1	12.3	15.1
2	11.8	14.8
3	12.5	15.5
4	12.0	14.9
5	11.6	14.7

a) Calculate the mean height increase for plants grown with standard fertiliser. [2 marks]

Calculate the sum of the individual height increases: $12.3 + 11.8 + 12.5 + 12.0 + 11.6 = 60.2$ [1 mark]

Divide this by the number of values to find the mean: $\dfrac{60.2}{5} = ? \ldots$ [1 mark]

b) Calculate the mean height increase for plants grown with the new fertiliser. [2 marks]

Calculate the sum of the individual height increases: ... [1 mark]

Divide this by the number of values to find the mean: ... [1 mark]

Q2. A student carried out a titration. The table shows the student's results.

	Trial 1	Trial 2	Trial 3	Trial 4
Volume of solution (cm^3)	23.80	22.60	22.65	22.55

Calculate the mean volume of solution.
Only use the concordant data (results within 0.10 cm^3 of each other). [3 marks]

Identify the concordant data: 22.60 cm^3, 22.65 cm^3, 22.55 cm^3 [1 mark]

Find the sum of the concordant data: $22.60 + 22.65 + 22.55 = 67.80$ cm^3 [1 mark]

Divide this by the number of values to find the mean: ... [1 mark]

Q3. A student carried out a titration. The table shows the student's results.

	Trial 1	Trial 2	Trial 3	Trial 4	Trial 5
Volume of solution (cm^3)	32.65	33.50	32.90	32.75	32.70

Calculate the mean volume of solution.
Only use the concordant data (results within 0.10 cm^3 of each other). [3 marks]

Identify the concordant data: 32.65 cm^3, 32.75 cm^3, 32.70 cm^3 [1 mark]

Find the sum of the concordant data: ... [1 mark]

Divide this by the number of values to find the mean: ... [1 mark]

Biology questions

Q1. A student recorded the height of four seedlings after seven days.
Calculate the arithmetic mean of these heights: 6.2 cm   5.8 cm   6.5 cm   6.3 cm

....[2 marks]

Q2. A scientist records a patient's pulse rate three times before exercise and after exercise.

Trial	Pulse before exercise (beats per minute)	Pulse after exercise (beats per minute)
1	72	126
2	70	128
3	74	130

a) Calculate the mean pulse rate before exercise. ....[2 marks]

b) Calculate the mean pulse rate after exercise. ....[2 marks]

Q3. The mass of five leaves from the same plant was recorded.
Calculate the mean leaf mass.

Leaf number	1	2	3	4	5
Mass (g)	0.42	0.39	0.41	0.40	0.35

....[2 marks]

Q4. A student was investigating the effect of light intensity on the rate of photosynthesis in a water plant. The number of oxygen bubbles produced in 1 minute at different distances from a lamp was recorded.

Calculate the mean number of bubbles produced at each distance.

Distance from lamp (cm)	Number of bubbles		
	Trial 1	Trial 2	Trial 3
10	27	28	26
30	12	14	13
50	3	1	5

....

....[6 marks]

Q5. A medical technician is studying the concentration of protein in eight patients' blood samples.

Patient number	1	2	3	4	5	6	7	8
Protein concentration (g/L)	72.4	70.8	75.2	68.9	71.5	124.8	73.2	69.7

a) Calculate the mean protein concentration using all eight samples. ....

....[2 marks]

b) Identify the anomalous result and recalculate the mean without it. ....

....[2 marks]

[Total marks 18]

Chemistry questions

Q1. The volume of gas produced in 60 seconds during an experiment was recorded.

	Trial 1	Trial 2	Trial 3
Volume of gas (cm³)	44	42	43

Calculate the mean volume of gas. ....

....[2 marks]

Q2. A student investigated the reaction between zinc and hydrochloric acid. They measured the temperature rise during the reaction.

	Trial 1	Trial 2	Trial 3	Trial 4
Increase in temperature (°C)	17.6	17.2	17.8	17.4

Calculate the mean increase in temperature for the reaction. ....

....[2 marks]

Q3. During an investigation, the decrease in mass was measured.

	Trial 1	Trial 2	Trial 3	Trial 4
Decrease in mass (g)	0.63	0.62	0.66	0.65

Calculate the mean decrease in mass. ....

....[2 marks]

Q4. A student carried out the decomposition of copper carbonate. They measured and recorded the mass of copper oxide (CuO) produced.

	Trial 1	Trial 2	Trial 3	Trial 4	Trial 5
Mass of CuO (g)	2.59	2.58	2.60	2.58	2.55

Calculate the mean mass of copper oxide. ....

....[2 marks]

Q5. The table below shows the results of a titration.

	Trial 1	Trial 2	Trial 3	Trial 4
Volume of solution (cm³)	24.20	24.45	24.55	24.50

Calculate the mean volume of solution.

Only use the concordant data (results within 0.10 cm³ of each other)..

....

....[2 marks]

[Total marks 10]

Arithmetic mean

Physics questions

Q1. The mean time for a pendulum to complete ten oscillations was measured.

	Trial 1	Trial 2	Trial 3	Trial 4
Time to complete ten oscillations (s)	15.2	15.8	15.5	15.4

Calculate the mean time for ten oscillations of the pendulum. ....

....[2 marks]

Q2. The current flowing through a circuit is measured at different times.

Time (minutes)	0	5	10	15
Current (A)	2.0	1.9	1.7	1.8

Calculate the mean current. ....

....[2 marks]

Q3. A light gate is used to measure the speed of a trolley as it moves down a ramp at the same height. The experiment was repeated five times.

	Trial 1	Trial 2	Trial 3	Trial 4	Trial 5
Speed (m/s)	0.45	0.42	0.46	0.43	0.44

Calculate the mean speed of the trolley. ....

....[2 marks]

Q4. An engineer is investigating the efficiency of different electric motors. The efficiency (%) is calculated for each motor in four trials.

Motor	Efficiency (%)			
	Trial 1	Trial 2	Trial 3	Trial 4
A	84	82	85	83
B	76	74	75	42
C	68	70	67	69

Identify any anomalous result(s) and calculate the mean efficiency for each motor.

....

....

....

....

....[6 marks]

[Total marks 12]

Sampling

If we wanted to know how many daisy plants were in a field, we could count them. However, that is time consuming, and we might make mistakes, so we use a technique called sampling. Sampling is a method of collecting data from a portion of a population and using it to make inferences about the whole population. Sampling must be random to avoid bias and the sample size should be as large as possible. Quadrats are often used to sample habitats. Quadrats come in different sizes, so it is important to know the size used.

Sampling is most frequently tested in biology. There are no questions in this section for chemistry or physics.

Worked examples

Q1. Quadrats are used to sample the population of clover in a field. The field measures 60 m × 40 m. Quadrats of 1 m × 1 m are placed randomly in 25 places and 150 clover plants are counted.

Estimate the population in the field. [5 marks]

Calculate the area of one quadrat:	1 m × 1 m = 1 m^2	[1 mark]
Calculate the total area of the field:	60 × 40 = 2400 m^2	[1 mark]
Calculate the total area sampled:	no. of quadrats × area of one quadrat = 25 × 1 = 25 m^2	
Calculate the density of clover:	density $= \dfrac{\text{total organisms counted}}{\text{total area sampled}} = \dfrac{150}{25}$	[1 mark]
	= 6 clover plants per m^2	[1 mark]
Estimate the population in the field:	population estimate = density × total area of field	
Substitute values into equation:	= 6 × 2400	
Calculate the answer:	6 × 2400 = 14 400 clover plants in the field	[1 mark]

Q2. A student uses a 0.5 m × 0.5 m quadrat to sample plant species in a wood. The quadrat is placed randomly ten times. The total number of snowdrops counted was 43.

Estimate the number of snowdrops per m^2. [4 marks]

Calculate the area of one quadrat:	0.5 × 0.5 = 0.25 m^2	[1 mark]
Calculate the total area sampled (number of quadrats placed × area of one quadrat):	10 × 0.25 m^2 = 2.5 m^2	[1 mark]
Write out the equation:	density $= \dfrac{\text{total organisms counted}}{\text{total area sampled}}$	
Substitute values into equation:	density $= \dfrac{43}{2.5}$	[1 mark]
Calculate the answer:	17.2 snowdrops per m^2	[1 mark]

Faded examples

Q1. Quadrats are used to sample the population of daisy plants in a field. The field measures 80 m × 30 m. Quadrats of 1 m × 1 m are placed randomly in 15 places and 30 daisy plants are counted.

Estimate the population in the field. [6 marks]

Calculate the area of one quadrat: $1 \text{ m} \times 1 \text{ m} = 1 \text{ m}^2$ [1 mark]

Calculate the total area of the field: $80 \times 30 = 2400 \text{ m}^2$ [1 mark]

Calculate the total area sampled: $15 \text{ quadrats} \times 1 \text{ m}^2 = 15 \text{ m}^2$ [1 mark]

Calculate the density of daisies: $\dfrac{\text{total organisms counted}}{\text{total area sampled}} = \dfrac{30}{15}$ [1 mark]

Estimate the plant population: $= 2 \text{ plants per m}^2$

population estimate = density × total area of field

Substitute values into equation: … [1 mark]

Calculate the answer: … [1 mark]

Q2. An ecologist is investigating the population of orchids in a meadow. The ecologist used a quadrat which measured 0.5 m × 0.5 m and was placed randomly in 20 locations across the meadow. In total, 45 orchids were found. Estimate the number of orchids per m². [4 marks]

Calculate the area of one quadrat: $0.5 \times 0.5 = 0.25 \text{ m}^2$ [1 mark]

Calculate the total area sampled: $20 \times 0.25 = …$. m² [1 mark]

Calculate the density: … [1 mark]

Write the answer: … [1 mark]

Biology questions

Q1. The school field measures 100 m × 80 m. A student uses a 1 m × 1 m quadrat to sample dandelions. The quadrat is randomly placed 20 times. The total number of dandelions counted is 85.

Estimate the total population of dandelions in the field. ....

....

....

....

....[5 marks]

Q2. Students placed twenty 1 m × 1 m quadrats on the playing field to count daisy plants. The playground measures 50 m × 30 m. Here are the results:

Quadrat	1	2	3	4	5	6	7	8	9	10	11	12	13	14	15	16	17	18	19	20
No. of daisies	4	7	0	8	2	8	7	2	9	3	8	5	3	0	9	4	2	7	3	6

Estimate the total population of daisies on the playing field. ....

....

....

....

....[5 marks]

Q3. A student investigates heather plants on a hillside area measuring 75 m × 45 m. They use 32 quadrats of 0.5 m × 0.5 m placed randomly and count a total of 96 heather plants.

Estimate the total population of heather in the hillside area. ....

....

....

....

....[5 marks]

Q4. A student uses a 0.5 m × 0.5 m quadrat to sample the plant species in a field. The quadrat is placed randomly 12 times. The total number of buttercups counted was 36.

Estimate the number of buttercups per m². ....

....

....

....[4 marks]

Sampling

Q5. An ecologist was concerned about the population of banded snails in two woodlands. The ecologist sampled each area with ten quadrats measuring 0.5 m × 0.5 m. The results are recorded below:

Quadrat	Woodland A: Area 472 m² Number of snails	Woodland B: Area 720 m² Number of snails
1	3	5
2	4	6
3	5	2
4	0	0
5	2	0
6	6	0
7	3	3
8	1	4
9	1	7
10	0	3

a) Calculate the area of the quadrat used.

.... [1 mark]

b) Calculate the total area sampled in woodland A.

.... [1 mark]

c) Calculate the number of snails per m² for woodland A and woodland B.

....

....[2 marks]

d) Estimate the total population of snails in woodland A and woodland B.

....

....[2 marks]

[Total marks 25]

Simple probability

In biology, we use simple probability to calculate the likelihood of inheriting specific traits or genotypes. This can be expressed using fractions, ratios, decimals or percentages. The probability of an outcome is a number between 0 (impossible) and 1 (certain). We can represent this probability as a decimal, fraction, percentage or ratio. In genetics, Punnett squares are used to help predict the probability and ratios of offspring inheriting particular traits, especially in monohybrid crosses. These questions require an understanding of the following terms: monohybrid cross, genotype, phenotype, heterozygous, homozygous, dominant and recessive. There are no questions in this section for chemistry or physis.

Worked examples

Q1. In a genetic cross between two heterozygous parents (Aa), what is the probability of producing an offspring with the genotype AA? [2 marks]

Is the question asking about genotype or phenotype?

genotype

Complete a Punnett square showing the cross: [1 mark]

Parent 1 gametes

Parent 2 gametes	A	a
A	AA	Aa
a	Aa	aa

Write down the equation:

$$\frac{\text{number of desired outcomes}}{\text{total outcomes}}$$

Substitute values into the equation:

$$\frac{1}{4}$$

Convert into a fraction, decimal, percentage or ratio as the answer:

The probability of producing an offspring with the genotype AA is 25% (accept ¼ or 0.25 or 1:3) [1 mark]

Q2. In chickens, feathered legs (F) is dominant to clean legs (f). A breeder wants to know the probability of getting a clean-legged chick from two birds that are heterozygous.
Express this probability as a decimal. [2 marks]

Is the question asking about genotype or phenotype?

phenotype

Complete a Punnett square showing the cross: [1 mark]

Parent 1 gametes

Parent 2 gametes	F	f
F	FF	Ff
f	Ff	ff

Write down the equation:

$$\frac{\text{number of desired outcomes}}{\text{total outcomes}}$$

Substitute values into the equation:

$$\frac{3}{4}$$

Convert into the answer:

The probability of producing an offspring with the phenotype clean legs is 0.75. [1 mark]

Faded examples

Q1. In tomatoes, red fruit (R) is dominant to yellow fruit (r).
What is the probability that an offspring from a heterozygous and a homozygous recessive parent plant will have yellow fruit? [2 marks]

Is the question asking about genotype or phenotype?   phenotype

Complete a Punnett square showing the cross: [1 mark]

Parent 1 gametes

	R	r
r	Rr	rr
r	Rr	rr

Parent 2 gametes

Write down the equation: $\dfrac{\text{number of desired outcomes}}{\text{total outcomes}}$

Substitute values into the equation: $\dfrac{2}{4}$

Convert into the answer: …   [1 mark]

Q2. In cattle, polled (no horns) heads (P) are dominant to horned heads (p). If two heterozygous cattle are crossed, what is the probability of offspring being polled?
Express the answer as a decimal. [2 marks]

Is the question asking about genotype or phenotype?   phenotype

Complete a Punnett square showing the cross: [1 mark]

Parent 1 gametes

	P	p
P	PP	Pp
p	Pp	pp

Parent 2 gametes

Write down the equation: $\dfrac{\text{number of desired outcomes}}{\text{total outcomes}}$

Substitute values into the equation: …

Convert into the answer: …   [1 mark]

Biology questions

Q1. In humans, each baby has an equal chance of being born as a male or a female. What is the probability that an offspring will be female?

.... [1 mark]

Q2. In a genetic cross between two heterozygous parents (Aa), what is the probability of producing offspring with the genotype aa?

Use the completed Punnett square to write down the probability as a percentage.

	A	**a**
A	AA	Aa
a	Aa	aa

.... [1 mark]

Q3. In humans, brown eyes (B) are dominant over blue eyes (b). If one parent is homozygous dominant (BB) and the other is homozygous recessive (bb), what is probability their child will have brown eyes?

Complete the Punnett square and write down the probability as a percentage.

	B	**B**
b		
b		

....

....[2 marks]

Q4. In a population of fruit flies, the allele for red eyes (R) is dominant to that for white eyes (r). If a heterozygous fly and a homozygous recessive fly are crossed, what is the probability of producing white-eyed flies?

Complete the Punnett square and express the answer as a decimal.

	R	**r**
r		
r		

....

....[2 marks]

Q5. In pea plants, tall (T) is dominant over short (t). If a homozygous tall plant is crossed with a homozygous short plant, what is the probability that an offspring will be tall?

.... [1 mark]

Q6. In humans the ability to roll the tongue is dominant (R) over the inability to roll (r) it. If two parents who are heterozygous for tongue rolling have offspring, what is the probability that a child will be able to roll the tongue?

Complete the Punnett square and write down the probability. [2 marks]

Q7. In snapdragon plants, tall stems (T) are dominant over short stems (t). A gardener crosses a homozygous tall plant with a heterozygous plant.

a) Complete the genetic diagram.

Parent alleles

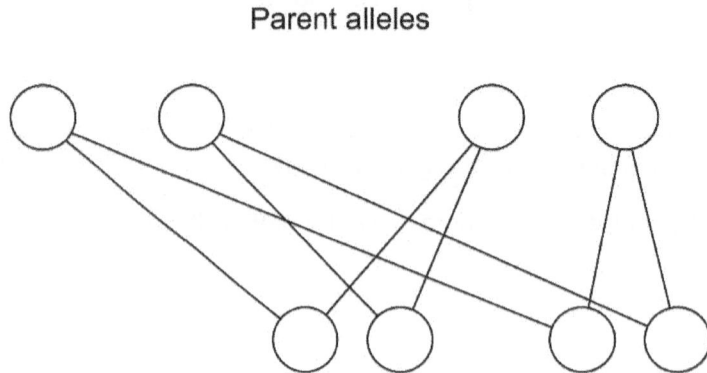

Offspring genotypes   —— ——   —— ——

Offspring phenotypes   —— ——   —— ——

[1 mark]

b) What is the probability as a percentage that the offspring will have short stems?

.... [1 mark]

Q8. In sheep, white wool (W) is dominant over black wool (w).
If two heterozygous sheep are crossed, what percentage of the offspring are likely to have black wool?

.... [1 mark]

Q9. In dogs, a curly tail (C) is dominant over a straight tail (c).
If Cc is crossed with cc, what fraction of the offspring will be homozygous recessive?

.... [1 mark]

Q10. In a genetics experiment, four pea plants are tall and two pea plants are short.
If you select one plant at random, what is the probability of selecting a tall plant? Give your answer as a fraction.

.... [1 mark]

Q11. In a population of fruit flies, twelve have red eyes and eight have white eyes.
What is the probability of randomly selecting a red eyed fruit fly?
Express this as: **a)** a ratio **b)** a fraction.

a) .... [1 mark]

b) .... [1 mark]

[Total marks 16]

Converting units

To convert between units, you need to use a conversion factor. For example, to convert grams to kilograms you need to divide by 1000. The 1000 is the conversion factor. To convert from a small unit, like grams, to a larger unit, like kilograms, you need to divide by the conversion factor. If you are converting from a larger unit to a smaller unit then you need to multiply by the conversion factor. If the units are squared then the conversion factor must be squared. In science you will need to convert between units with the common prefixes: nano-, micro-, milli-, kilo-, mega-, giga- and tera-. An example is shown below for the unit watt.

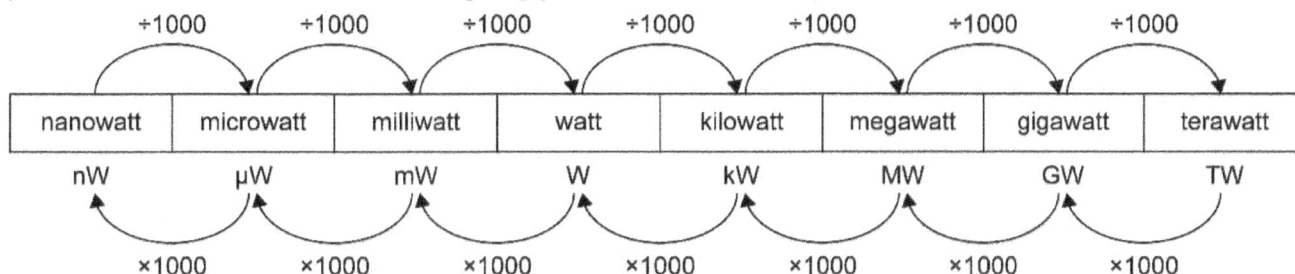

÷1000	÷1000	÷1000	÷1000	÷1000	÷1000	÷1000

nanowatt	microwatt	milliwatt	watt	kilowatt	megawatt	gigawatt	terawatt
nW	µW	mW	W	kW	MW	GW	TW

×1000	×1000	×1000	×1000	×1000	×1000	×1000

When dealing with liquids we also use the unit dm^3 (decimetre cubed). The conversion factor for cm to dm is 10. When the unit is cubed then the conversion factor is also cubed. So, the conversion factor for cm^3 to dm^3 is $10^3 = 1000$. When dealing with areas, we are working with squared units, which means the conversion factor must be squared too. $1\ m = 100\ cm$ but $1\ m^2$ converted to cm^2 is $(100\ cm)^2 = 100 \times 100 = 10\ 000\ cm^2$.

Worked examples

Q1. Convert 2.5 km to m. [1 mark]

Identify the conversion factor:	1000	
Is the unit getting larger or smaller?	Smaller. This means we multiply by the conversion factor.	
Calculate answer including units:	2.5 km × 1000 = 2500 m	[1 mark]

Q2. Convert 4.5 mm to m. [1 mark]

Identify the conversion factor:	1000	
Is the unit getting larger or smaller?	Larger. This means we divide by the conversion factor.	
Calculate answer including units:	4.5 mm ÷ 1000 = 0.0045 m	[1 mark]

Q3. Convert 25 cm^3 to dm^3. [1 mark]

Identify the conversion factor:	1000	
Is the unit getting larger or smaller?	Larger. This means we divide by the conversion factor.	
Calculate answer including units:	25 cm^3 ÷ 1000 = 0.025 dm^3	[1 mark]

Faded examples

Q1. Convert 1500 g to kg. [1 mark]

Identify the conversion factor: 1000

Is the unit getting larger or smaller? Larger. This means we divide by the conversion factor.

Calculate the answer including units: … [1 mark]

Q2. Convert 1.5 mg to g. [1 mark]

Identify the conversion factor: 1000

Is the unit getting larger or smaller? …

Calculate the answer including units: … [1 mark]

Q3. Convert 125 cm^3 to dm^3. [1 mark]

Identify the conversion factor: 1000

Is the unit getting larger or smaller? Larger. This means we divide by the conversion factor.

Calculate the answer including units: … [1 mark]

Q4. Convert 0.5 dm^3 to cm^3. [1 mark]

Identify the conversion factor: 1000

Is the unit getting larger or smaller? …

Calculate the answer including units: … [1 mark]

Biology questions

Q1. The diameter of a red blood cell is 7.5 micrometres (μm). Convert this to nanometres (nm).

.... [1 mark]

Q2. A bacterial cell measures 2000 nm in length. Convert this into micrometres (μm).

.... [1 mark]

Q3. Convert the following from μm to nm:

a) 4.3 μm   **b)** 27 μm   **c)** 0.375 μm   **d)** 120 μm   [1 mark each]

a)..   **c)**..

b)..   **d)**..

Q4. Convert the following from μm to mm:

a) 125 μm   **b)** 7.5 μm   **c)** 0.5 μm   **d)** 59 μm   [1 mark each]

a)..   **c)**..

b)..   **d)**..

Q5. A microscope has a resolution of 200 nm.
Convert this into **a)** micrometres, **b)** millimetres and **c)** centimetres.   [1 mark each]

a)..   **c)**..

b)..

Q6. The diameter of a human egg cell is 0.12 mm.
Express this in **a)** micrometres and **b)** nanometres.   [1 mark each]

a)..   **b)**..

Q7. The length of ten bacterial cells was measured under a microscope. The total length was 25 μm. If each bacterium is the same size, what is the length of one cell in nanometres?

.... [1 mark]

Q8. The lung capacity of an adult human is approximately 6.2 dm^3.
Convert this to cm^3.

.... [1 mark]

Q9. A quadrat has an area of 0.25 m^2.
What is the quadrat area expressed in cm^2?

.... [1 mark]

Q10. A leaf has a surface area of 850 cm^2.
Convert this area into m^2.

.... [1 mark]

[Total marks 19]

Chemistry questions

Q1. The diameter of a nanoparticle is 55 nm.
Convert this length to micrometres (μm).

.... [1 mark]

Q2. During an experiment, a student added 25.30 cm^3 of hydrochloric acid to a conical flask.
Convert this volume to dm^3.

.... [1 mark]

Q3. Convert the following from kg to g:

a) 2 kg   **b)** 40 kg   **c)** 3.2 kg   **d)** 0.35 kg   [1 mark each]

a)..   c)..

b)..   d)..

Q4. Convert the following from J to kJ:

a) 6000 J   **b)** 5200 J   **c)** 10.3 J   **d)** 0.5 J   [1 mark each]

a)..   c)..

b)..   d)..

Q5. Convert the following:

a) 5 g to mg   **b)** 0.4 mg to g   **c)** 100 mg to g   **d)** 0.01 g to mg   [1 mark each]

a)..   c)..

b)..   d)..

Q6. Convert the following:

a) 29.6 cm^3 to dm^3 **b)** 0.2 dm^3 to cm^3 [1 mark each]

a)..   b)..

Q7. A student determined the reacting volumes of hydrochloric acid and sodium hydroxide solution by titration. They measured 25 cm^3 of sodium hydroxide solution and poured it into a conical flask. Convert the volume of sodium hydroxide to dm^3.

.... [1 mark]

[Total marks 17]

Physics questions

Q1. A wire has a length of 250 cm.
Convert this into metres.

.... [1 mark]

Q2. A student measures the length of a ramp as 1.37 m.
Convert this into cm.

.... [1 mark]

Q3. Convert the following from mA to A:

a) 2500 mA   **b)** 1050 mA   **c)** 730 mA   **d)** 15 000 mA   [1 mark each]

a)..   **c)**..

b)..   **d)**..

Q4. Convert the following from $k\Omega$ to Ω:

a) 5 $k\Omega$ **b)** 79 $k\Omega$ **c)** 2.6 $k\Omega$ **d)** 0.5 $k\Omega$ [1 mark each]

a)..   **c)**..

b)..   **d)**..

Q5. Convert the following:

a) 25 cm into μm   **b)** 0.9 mg to g   **c)** 102 m into km   **d)** 21.8 km into m   [1 mark each]

a)..   **c)**..

b)..   **d)**..

Q6. Convert the following:

a) 378 cm^3 to dm^3 **b)** 0.73 dm^3 to cm^3 **c)** 1 MHz to Hz   **d)** 2 TW to W   [1 mark each]

a)..   **c)**..

b)..   **d)**..

[Total marks 18]

Chapter 3 Algebra

Solving equations

Solving equation questions in science often requires different steps and skills. You will need to be comfortable with changing the subject of an equation, combining and cancelling like terms, and substituting known values to find an unknown one. To maximise chances of getting marks, it is vital that you show clear working out at all times while carrying out calculation questions. To do this, we recommend substituting the values into the equation before rearranging. In some questions you may need to convert units before carrying out the calculation. However, in these exercises you will not need to convert units.

How to change the subject of an equation

When carrying out a calculation involving equations you may need to rearrange the equation. In mathematics. you learn how to change the subject of an equation. In science, you may also need to do this in order to carry out a calculation. An example question is given below showing how to change the subject of an equation.

Q1. The size of a real cell is 40 μm. The cell was magnified ×200.
Calculate the size of the image. [3 marks]

Write down the equation:
$$\text{magnification} = \frac{\text{size of image}}{\text{size of real object}}$$

Substitute values into equation:
$$200 = \frac{\text{size of image}}{40 \text{ μm}}$$
[1 mark]

Change the subject of the equation to size of image:

We need to have 'size of image' by itself on one side of the equation. To do this we multiply both sides of the equation by 40 μm:
$$200 \times 40 \text{ μm} = \frac{\text{size of image}}{40 \text{ μm}} \times 40 \text{ μm}$$

Which can be written as:
$$200 \times 40 \text{ μm} = \frac{\text{size of image} \times 40 \text{ μm}}{40 \text{ μm}}$$

Now we simplify the equation. On the right-hand side we are both multiplying and dividing by 40 μm so they cancel each other out (40 ÷ 40 = 1).
$$200 \times 40 \text{ μm} = \frac{\text{size of image} \times \cancel{40} \text{ μm}}{\cancel{40} \text{ μm}}$$

This leaves us with: 200 × 40 μm = size of image

Calculate answer:
$$200 \times 40 = 8000$$
[1 mark]

Write units:
$$8000 \text{ μm}$$
[1 mark]

The following question focuses on changing the subject of an equation without substitution.

Q2. Change the subject of the following equation to frequency:

$$\text{wave speed} = \text{frequency} \times \text{wavelength}$$ [1 mark]

Write down the equation:

wave speed = frequency × wavelength

Divide both sides by 'wavelength':

$$\frac{\text{wave speed}}{\text{wavelength}} = \frac{\text{frequency} \times \text{wavelength}}{\text{wavelength}}$$

Simplify the equation:

$$\frac{\text{wave speed}}{\text{wavelength}} = \text{frequency}$$ [1 mark]

Worked examples

Q1. The length of a cell in an image was 48 000 μm. The actual length of the cell was 120.
Calculate the magnification. [3 marks]

Write down the equation:

$$\text{magnification} = \frac{\text{size of image}}{\text{size of real object}}$$

Substitute values into equation:

$$\text{magnification} = \frac{48\ 000\ \mu m}{120\ \mu m}$$ [1 mark]

Change the subject of the equation (if needed):

No need to rearrange the equation to change the subject

Calculate answer:

$$48\ 000 \div 120 = 400$$ [1 mark]

Write units:

×400 [1 mark]

(magnification has no unit, instead × is placed before the number)

Q2. The resistance of a component is 0.60 Ω. The potential difference across the component is 0.45 V.

Calculate the current in the circuit. [3 marks]

Write down the equation:

potential difference = current × resistance

Substitute values into equation:

0.45 V = current × 0.60 Ω [1 mark]

Change the subject of the equation (if needed):

Make current the subject of the equation:

$$\text{current} = \frac{0.45\ V}{0.60\ \Omega}$$

Calculate answer:

$$0.45 \div 0.60 = 0.75$$ [1 mark]

Write units:

0.75 A [1 mark]

Solving equations

Faded examples

Q1. Change the subject of the following equation to time:

charge flow = current × time

[1 mark]

Write down the equation:

charge flow = current × time

Divide both sides of the equation by current:

$$\frac{charge\ flow}{current} = \frac{current \times time}{current}$$

Simplify the equation: …   ..   [1 mark]

Q2. Change the subject of the following equation to mass: $density = \frac{mass}{volume}$ [1 mark]

Write down the equation: $density = \frac{mass}{volume}$

Multiply both sides of the equation by volume: …   ..

Simplify the equation: …   ..   [1 mark]

Q3. Use the equation below to find the current [I] when power [P] is 60 W and the potential difference is 12 V.

power = potential difference × current

[4 marks]

Write down the equation: power = potential difference × current

Substitute values into equation: Substitute 60 W = 12 V × current [A]   [1 mark]

Change the subject of the equation (if needed): $\frac{60\ W}{12\ V} = \frac{12\ V \times current}{12\ V}$ [1 mark]

Calculate answer: …   ..   [1 mark]

Write units: …   ..   [1 mark]

Q4. Use the equation below to calculate the height in metres when mass is 5 kg, gravitational field strength is 9.8 N/kg and gravitational potential energy is 490 J.

gravitational potential energy = $m \times g \times h$

[4 marks]

Write down the equation: gravitational potential energy = $m \times g \times h$

Substitute values into equation: 490 J = 5 kg × 9.8 N/kg × h [1 mark]

Change the subject of the equation (if needed):   …   ..   [1 mark]

Calculate answer: …   ..   [1 mark]

Write units: …   ..   [1 mark]

Q5. An object was travelling at a velocity of 10 m/s. The object had 32 000 J of kinetic energy.
Calculate the mass of the object in kg. [4 marks]

Write down the equation: kinetic energy = 0.5 mv^2

Substitute values into equation: $32\ 000 = 0.5 \times m \times 10^2$ [1 mark]

Change the subject of the equation (if needed):

$$\frac{32\ 000\ \text{J}}{0.5 \times 10^2} = \frac{0.5 \times m \times 10^2}{0.5 \times 10^2}$$ [1 mark]

$$\frac{32\ 000\ \text{J}}{0.5 \times 10^2} = m$$

Calculate answer: ...   .. [1 mark]

Write units: ...   .. [1 mark]

Q6. An object was travelling at a velocity of 5 m/s. The object had 7900 J of kinetic energy.
Calculate the mass of the object in kg. [4 marks]

Write down the equation: kinetic energy = 0.5 mv^2

Substitute values into equation: $7900 = 0.5 \times m \times 5^2$ [1 mark]

Change the subject of the equation (if needed):   ...   .. [1 mark]

Calculate answer: ...   .. [1 mark]

Write units: ...   .. [1 mark]

Biology questions

For Questions 1, 3, 4, 5 and 7 use the equation: **a)** magnification = $\dfrac{\text{size of image}}{\text{size of real object}}$

For Questions 6 and 8 use the equation:

b) estimated population = $\dfrac{\text{average number of organisms in a quadrat}}{\text{area of quadrat}}$ × total area

Q1.   Change the subject of equation **a)** to 'size of image'.

.... [1 mark]

Q2.   Change the subject of the following equation to 'eyepiece lens magnification':

total magnification = eyepiece lens magnification × objective lens magnification

.... [1 mark]

Q3.   A magnified image of a cell is 2 mm long. The actual length of the cell is 0.02 mm.

Calculate the magnification used..

....

....[3 marks]

Q4.   A magnified image of a palisade cell is 15 000 µm wide. The actual width of the cell is 30 µm.

Calculate the magnification used..

....

....[3 marks]

Q5.   A microscope has a magnification of ×400. A student measures an image of an organelle
to be 8 mm long.
Calculate the real size of the organelle. ....

....

....[3 marks]

Q6.   In a quadrat study, a student finds 8 dandelions in a 1 m² quadrat. The total area studied is 200 m².

Estimate the total population of dandelions in the area. ....

....

....[3 marks]

Q7.   An image of a chloroplast is magnified ×400. It measures 20 mm in length.

Calculate the real size of the chloroplast. ....

....[3 marks]

Q8.   A student counts 6 clover plants in a 0.5 m² quadrat. The total area studied is 150 m².
Estimate the total population of clover plants in the area.

....

....[3 marks]

[Total marks 20]

Chemistry questions

For Questions 1, 3, and 5 use the equation: **a)** $R_f = \dfrac{\text{distance moved by substance}}{\text{distance moved by solvent}}$

For Questions 2, 4 and 6 use the equation: **b)** concentration $= \dfrac{\text{mass}}{\text{volume}}$

For Question 7 and 8 use the equation: **c)** amount of substance (mol) $= \dfrac{\text{mass of substance (g)}}{\text{relative formula mass (g/mol)}}$

Q1. Change the subject of equation **a)** to 'distance moved by substance'.

.... [1 mark]

Q2. Change the subject of equation **b)** to 'mass'.

.... [1 mark]

Q3. A student investigated a dye using chromatography. The distance moved by a spot was 3.4 cm. The solvent front moved 5.0 cm. Calculate the R_f value for the dye using equation **a)**.

....

....[3 marks]

Q4. 2.5 g of sodium chloride was dissolved in 50 cm^3 of water.
Calculate the concentration of the solution using equation **b)**.

....

....[3 marks]

Q5. A student investigated a dye using chromatography. The distance moved by the solvent was 12.4 cm. The R_f value was 0.5. Calculate the distance travelled by the substance using equation **a)**.

....

....[3 marks]

Q6. What mass of copper sulfate needs to be added to 250 cm^3 of water to form a 0.27 g/cm^3 solution?

Use equation **b)** for your calculation. ....

....

....[3 marks]

Q7. **HT** What mass of carbon contains 0.05 moles of carbon atoms?
The relative atomic mass of carbon is 12.

Calculate the mass using equation **c)**. ....

....

....[3 marks]

Q8. **HT** What mass of water contains 0.35 moles of water molecules?
The relative formula mass of water is 18.

Calculate the mass using equation **c)**. ....

....

....[3 marks]

[Total marks 20]

Physics questions

Q1. Change the subject of the following equation to 'mass':   resultant force = mass × acceleration

.... [1 mark]

Q2. Change the subject of the following equation to 'speed':   distance travelled = speed × time

.... [1 mark]

Q3. A sound wave has a frequency of 50 Hz. Calculate the period of the wave using the equation below:

$$\text{period (s)} = \frac{1}{\text{frequency (Hz)}}$$ ....

....

.... [3 marks]

Q4. A crane lifts a load vertically through a distance of 8 m. The work done by the crane is 200 J.

Calculate the force applied by the crane using the equation: work done = force × distance

....

.... [3 marks]

Q5. A kettle has a power of 3000 W. Calculate the time taken to do 30 000 J of work.

Use the equation: $\text{power (W)} = \frac{\text{work done (J)}}{\text{time (s)}}$ ....

....

.... [3 marks]

Q6. A student measures the current through a resistor as 3 A and the potential difference across the resistor is 12 V.
Calculate the resistance of the resistor. Use the equation: $V = IR$. ....

....

.... [3 marks]

Q7. A motor has an efficiency of 0.75. The useful power output of the motor is 150 W.

Calculate the total input to the motor. Use the equation: $\text{efficiency} = \frac{\text{useful power output}}{\text{total power input}}$

....

.... [3 marks]

Q8. A cyclist starts from rest and accelerates at 2 m/s^2.

Calculate the velocity of the cyclist after travelling 28 m.

Use the equation: (final velocity)^2 − (initial velocity)^2 = 2 × acceleration × distance

....

....

.... [3 marks]

[Total marks 20]

Chapter 4 Graphs

Interpreting line graphs

*Graphs and charts help us identify patterns and trends in data. A line of best fit can be used to find the value of y for any value of x (or vice versa). This is known as **interpolation**. Graphs can be used to make predictions about data outside the known values. This is known as **extrapolation**.*

To describe trends (or patterns) shown by a graph, you will need to describe the whole trend. This is important if the line of best fit is a curve. You should also use data in your description if possible.

- *Some graphs show a variable on the y-axis and time on the x-axis. This is known as a **time series** graph. In a time series graph, you need to describe how the variable changes over time.*

The description can also include the following terms to describe the relationship between variables:

- ***linear** – the line of best fit forms a straight line.*
- ***directly proportional** – as one variable doubles the other variable doubles. This forms a straight line that passes through the origin (0,0). If the line does not pass through the origin it is a proportional relationship, but not directly proportional.*
- ***inversely proportional** – as one variable doubles the other variable halves. This forms a curve.*

Worked examples

Q1. A student monitored the rate of a reaction by collecting hydrogen gas.

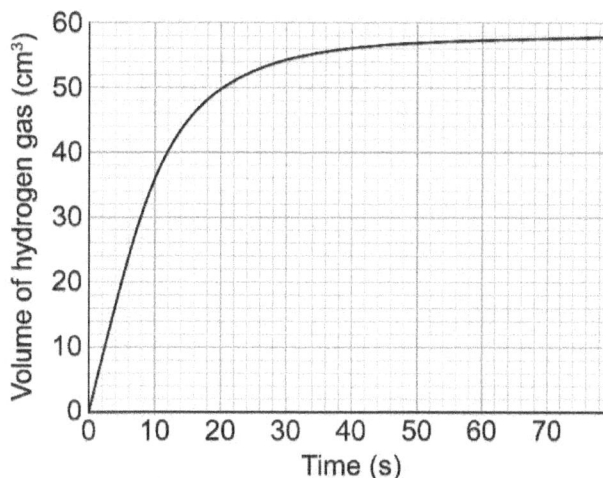

a) What volume of gas was produced after 30 seconds?   [1 mark]

Find the value on the appropriate axis:	30 seconds is found on the x-axis
Draw a straight line from this axis to the line of best fit. Draw a second straight line from the line of best fit to the other axis:	Draw a vertical line up from 30 seconds on the x-axis to the line of best fit. At the point these lines meet, draw a horizontal line to the y-axis.
Read the value from the other axis (with units):	54 cm^3

[1 mark]

b) How long did it take to produce 30 cm^3 of hydrogen gas?   [1 mark]

Find the value on the appropriate axis:	30 cm^3 is found on the y-axis
Draw a straight line from this axis to the line of best fit. Draw a second straight line from the line of best fit to the other axis:	Draw a horizontal line from 30 cm^3 to the line of best fit. At the point these lines meet, draw a horizontal line to the x-axis.
Read the value from the other axis (with units):	8 s

[1 mark]

c) Describe how the rate of reaction changes over time. [3 marks]

Describe how the variable changes over time:	The line of best fit is a curve. This means we need to describe how the rate differs at different points on the graph. We need to include data to help the description:	
	The rate of reaction is fastest at the start.	[1 mark]
	After 10 seconds, the rate of reaction decreases.	[1 mark]
	The reaction stops after approximately 70 seconds.	[1 mark]

Q2. The graph below shows how the volume of a gas changes with temperature.

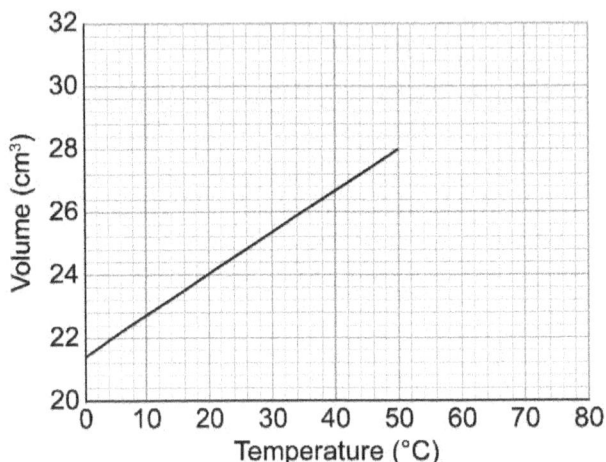

a) Predict the volume of gas in cm^3 at 80 °C. [1 mark]

Extend the line of best fit until it reaches the value given in the question:	Extend the line of best fit to 80 °C.	
Read the value using the line of best fit:	32 cm^3	[1 mark]

b) Describe any patterns or trends shown by the graph. [2 marks]

Describe how the variable on the x-axis affects the variable on the y-axis.	This graph has temperature (independent variable) plotted against volume of gas (dependent variable). This means we can use the phrase: 'As the [independent variable] increases, the [dependent variable] increases/decreases.'	
	As temperature increases, the volume of gas increases.	[1 mark]
Identify whether the relationship is linear, directly proportional or inversely proportional:	The line of best fit is straight; when extended it does not go through the origin (0,0), so we can write:	
	The relationship between the variables is linear.	[1 mark]

Faded examples

Q1. This graph shows the change in velocity of a car over time.

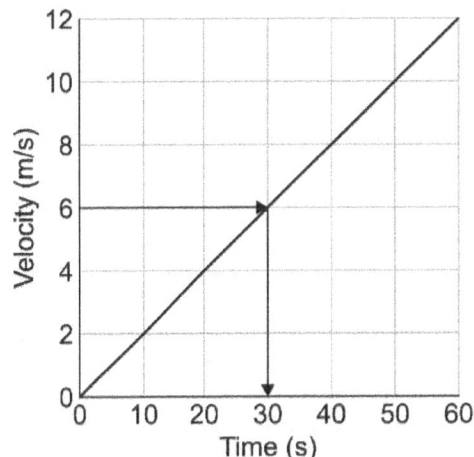

a) At what time did the car reach 6 m/s? [1 mark]

Find the value on the appropriate axis: 6 m/s is found on the y-axis.

Draw a straight line from this axis to the line of best fit. Draw a second straight line from the line of best fit to the other axis: See arrows on the graph.

Read the value from the other axis (with units): … [1 mark]

b) What was the velocity of the car after 50 seconds? [1 mark]

Find the value on the appropriate axis: 50 seconds is found on the x-axis.

Draw a straight line from this axis to the line of best fit. Draw a second straight line from the line of best fit to the other axis: Draw arrows on the graph.

Read the value from the other axis (with units): … [1 mark]

Q2. A student monitored the rate of a reaction over time.

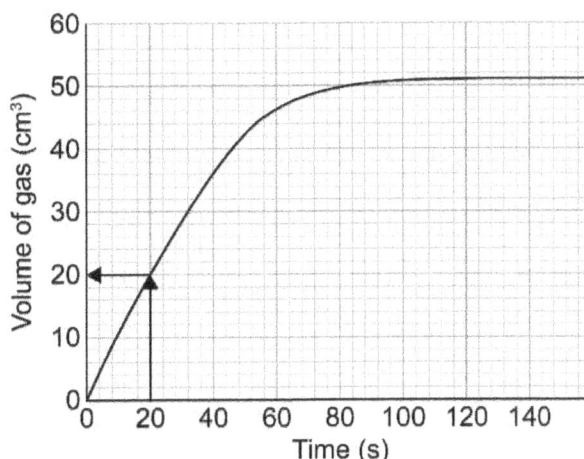

a) What volume of gas was produced after 20 seconds? [1 mark]

Find the value on the appropriate axis: 20 seconds is found on the x-axis.

Draw a straight line from this axis to the line of best fit. Draw a second straight line from the line of best fit to the other axis: See arrows on the graph.

Read the value from the other axis (with units): … [1 mark]

Interpreting line graphs

b) How long did it take to produce 42 cm^3 of gas? [1 mark]

Find the value on the appropriate axis: ...

Draw a straight line from this axis to the line of
best fit. Draw a second straight line from the
line of best fit to the other axis: Draw arrows on the graph.

Read the value from the other axis (with units): ... [1 mark]

Q3. A student investigated the activity of an enzyme at different temperatures. The graph shows the concentration of the product of the reaction at 20 °C.

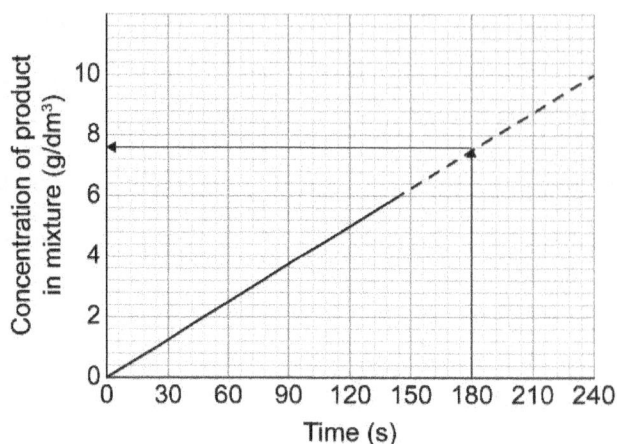

Predict the concentration of product in the mixture after 180 seconds. [1 mark]

Extend the line of best fit until it reaches
the value given in the question: Extend the line of best fit to 180 seconds.

Read the value using the line of best fit: ... [1 mark]

Q4. A student investigated the rate of transpiration by measuring the water uptake by a plant.

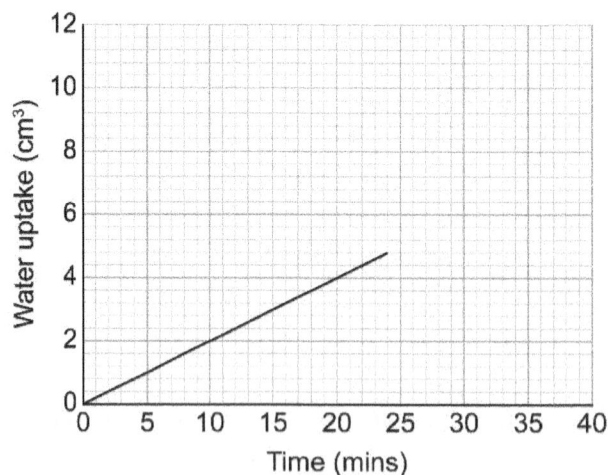

Predict the water uptake after 30 minutes. [1 mark]

Extend the line of best fit until it reaches
the value given in the question: Extend the line of best fit to 30 minutes.

Read the value using the line of best fit: ... [1 mark]

Biology questions

Q1. A student investigated the rate of photosynthesis by counting the number of oxygen bubbles produced by pondweed over time.

   a) How many bubbles were produced after 2 minutes?

   .... [1 mark]

   b) When did the rate of photosynthesis become constant?

   .... [1 mark]

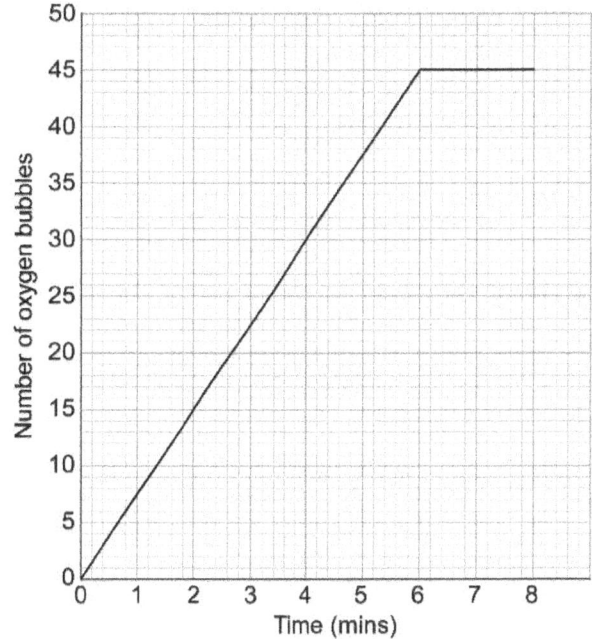

Q2. A student carried out an experiment to find out the effect of temperature on the rate of respiration in yeast by measuring the volume of carbon dioxide produced in 10 minutes.

   a) What volume of carbon dioxide was produced after 10 minutes at 30 °C?

   ....[1 mark]

   b) What volume of carbon dioxide was produced after 10 minutes at 60 °C?

   ....[1 mark]

   c) Describe how temperature affects the rate of respiration in yeast...

   ....[2 marks]

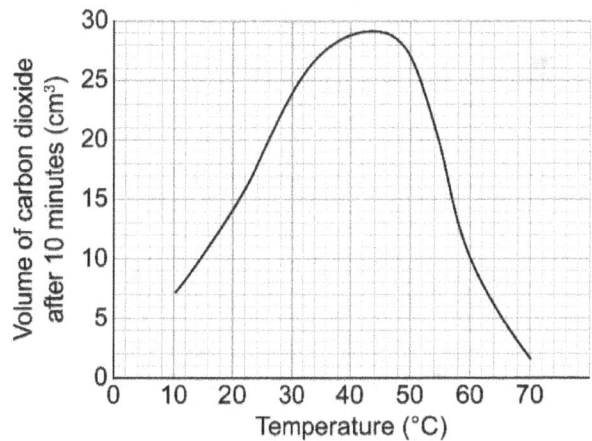

Q3. A student investigated the effect of pH on the rate of protein digestion by an enzyme. The student measured the time taken for the enzyme to digest 5 g of protein at different pH values.

   a) How long did it take to digest the protein at pH 3.0?

   ....[1 mark]

   b) At which pH value was the rate of digestion fastest?

   ....[1 mark]

   c) Describe how pH affects the rate of protein digestion by the enzyme. ....

   ....[2 marks]

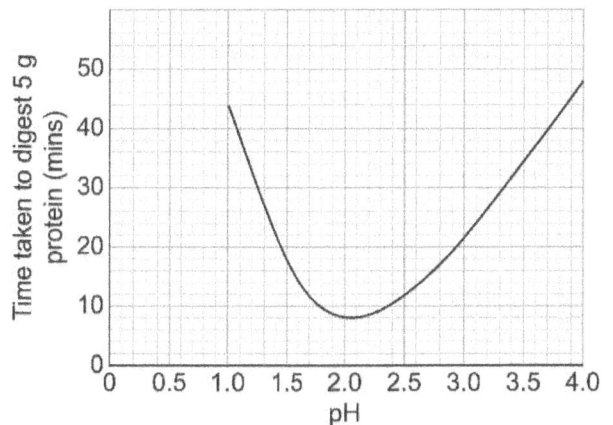

Interpreting line graphs

Q4. A student investigated the effect of
substrate concentration on enzyme activity
using the enzyme catalase and hydrogen
peroxide (H_2O_2).
The volume of oxygen released
was collected.

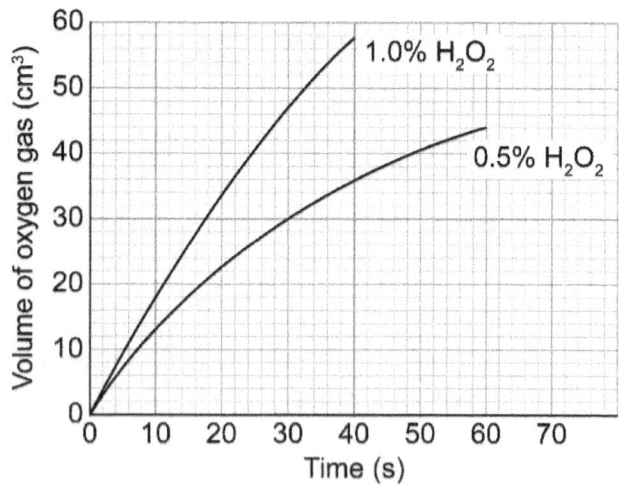

a) What volume of oxygen was produced
after 30 seconds using 1.0% H_2O_2?

....[1 mark]

b) How long did it take to produce 40 cm^3
of oxygen using 0.5% H_2O_2?

....[1 mark]

Q5. The graph shows the relationship between predator and prey populations over time.
Describe the pattern seen for both the predator and prey populations over time.

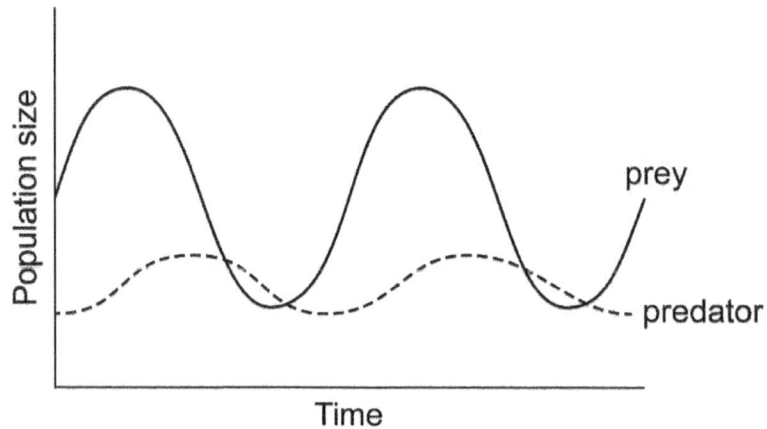

....

....

....[3 marks]

[Total marks 15]

Chemistry questions

Q1. A student monitored the volume of gas produced during an experiment.

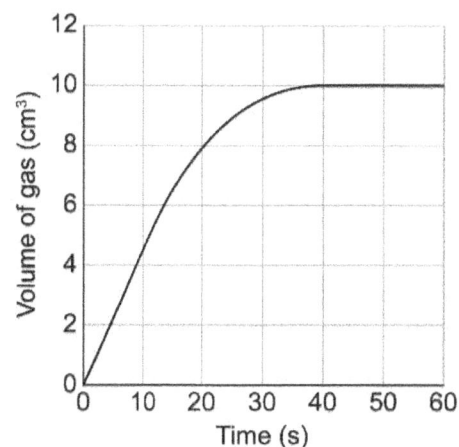

a) What volume of gas was produced after 20 seconds?

.... [1 mark]

b) When did the reaction stop?

.... [1 mark]

Q2. A student investigated the reaction between small marble chips and hydrochloric acid.

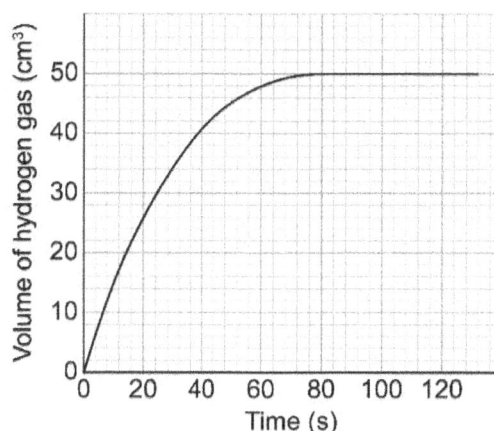

a) What volume of gas was produced after 20 seconds?

.... [1 mark]

b) How long did it take to produce 48 cm^3 of hydrogen gas?

.... [1 mark]

c) Describe how the rate of reaction changes over time.

....

....

....[3 marks]

Q3. A student investigated the effect of temperature on the reaction between magnesium and hydrochloric acid.

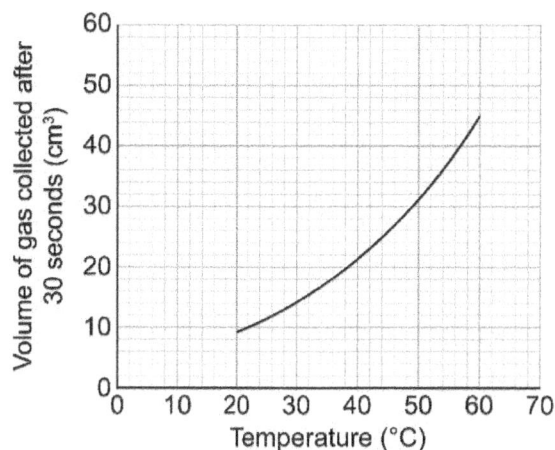

a) Predict the volume of gas collected after 30 seconds at 10 °C.

....[1 mark]

b) Describe how temperature affects the rate of reaction.

....

....

....[3 marks]

© HarperCollins*Publishers* Limited 2025

Interpreting line graphs

Q4. This graph shows how the yield of ammonia changes with temperature and pressure.

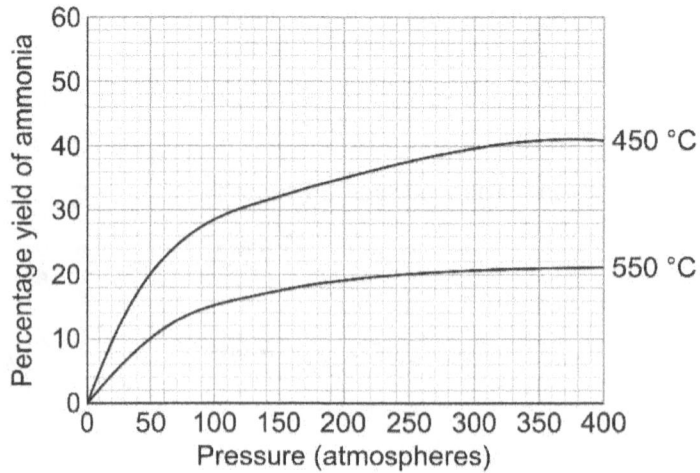

a) What is the percentage yield of ammonia at 50 atmospheres and 550 °C? .... [1 mark]

b) How does pressure affect the percentage yield of ammonia at 550 °C?

....

....[2 marks]

c) How does temperature affect the percentage yield of ammonia?

....[1 marks]

[Total marks 15]

Physics questions

Q1. The graph shows the velocity-time graph for a moving object.

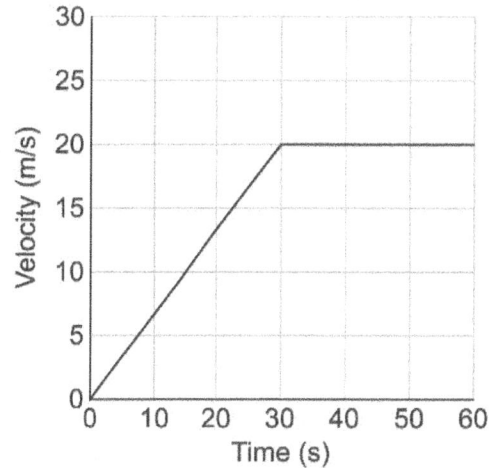

a) Find the velocity of the object at 10 seconds.

....[1 mark]

b) Find the velocity at 40 seconds.

....[1 mark]

c) How long does it take for the object to reach a velocity of 10 m/s?

....[1 mark]

d) Describe the relationship between velocity and time shown in the graph. ....

....

....[3 marks]

Q2. This is a velocity-time graph.

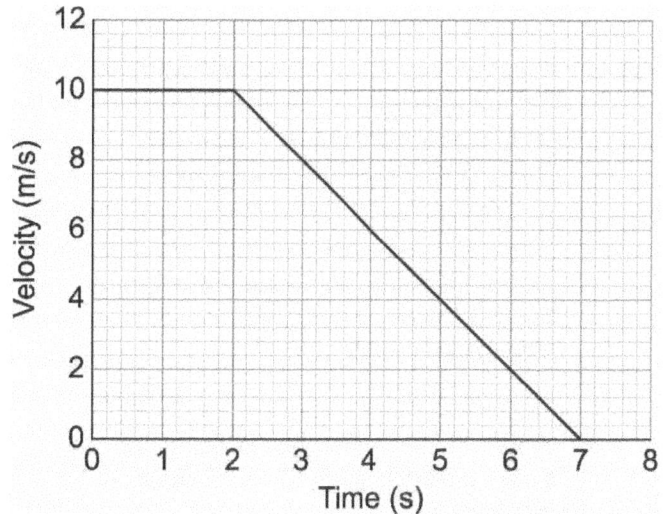

a) What is the velocity of the object at 4.0 seconds?

.... [1 mark]

b) When is the velocity of the object 0 m/s?

.... [1 mark]

c) Describe the trend seen in this graph. ....

....

....[3 marks]

Interpreting line graphs

Q3. A student investigated the extension of a spring when different forces are applied.
The results can be seen in the graph.

a) Find the extension of the spring when a force of 0.3 N is applied.

....[1 mark]

b) Predict the extension of the spring when a force of 0.1 N is applied.

....[1 mark]

c) Describe the relationship between the applied force and the extension of the spring.

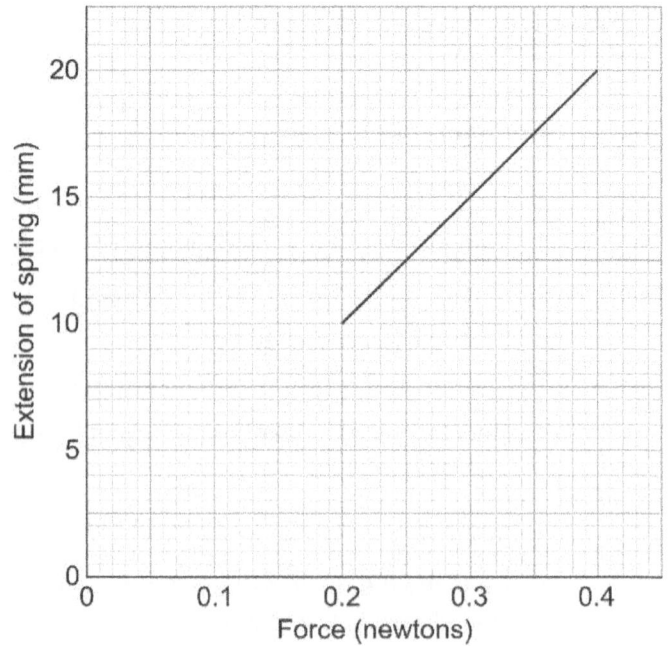

....

....

....[3 marks]

Q4. A student carried out an investigation to find out how current varies with potential difference for two different resistors.
The results are shown in the graph.

a) Use the graph to find the current through the 5 Ω resistor when the potential difference is 8.0 V.

....[1 mark]

b) What potential difference is being used if the current through the 10 Ω resistor is 1.2 A?

....[1 mark]

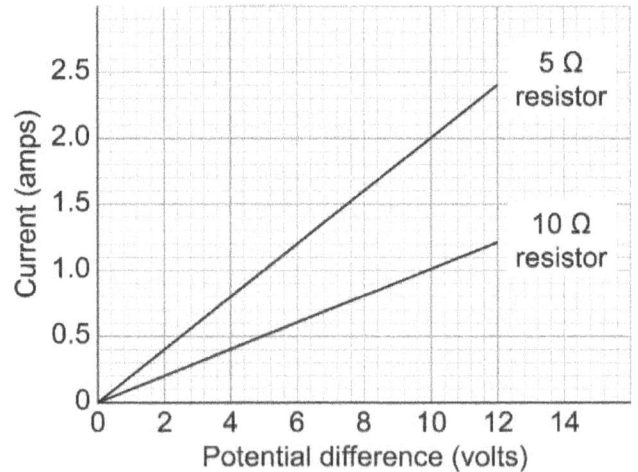

c) Describe the relationship between current and potential difference for the 5 Ω resistor.

....

....[2 marks]

d) Assuming the pattern continues, predict the current through the 10 Ω resistor at 14 V.

....[1 mark]

[Total marks 21]

Gradients and intercepts

The gradient of a line is also known as the slope. The gradient of a straight line is the same all the way along the line. Gradients are often used to show the 'rate' at which something is happening. The steeper the slope or gradient, the faster the rate of change.

The gradient can be found using the equation: $\quad gradient = \dfrac{change\ in\ y}{change\ in\ x} = \dfrac{\Delta y}{\Delta x}$

The mean rate of reaction between two points on a graph can be found using this equation, whether the line of best fit is straight or slightly curved.

The intercept is the point at which the line of best fit crosses an axis. The term often refers to the y-axis but can also apply to the x-axis. In the equation for a linear relationship, $y = mx + c$, the letter c refers to where the line crosses the y-axis, known as the y-intercept. The letter m refers to the gradient or slope.

Worked examples

Q1. A student investigated the rate of reaction between magnesium and dilute hydrochloric acid.

Determine the rate of reaction over the first 3 minutes by finding the gradient of the line. [4 marks]

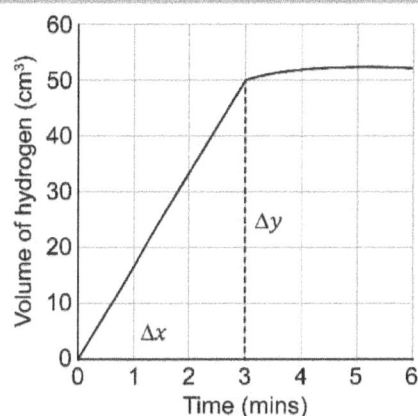

Find values for Δy and Δx:
$\Delta y = 50\ cm^3 - 0\ cm^3 = 50\ cm^3$ [1 mark]

$\Delta x = 3\ min - 0\ min = 3\ min$

Substitute values into equation:
$gradient = \dfrac{\Delta y}{\Delta x}$ [1 mark]

$gradient = \dfrac{50}{3}$

Calculate answer:
$\dfrac{50}{3} = 16.7$ [1 mark]

Write units:
Identify the units using information from the labels on [1 mark] the scales and the equation for the gradient. You do not need to show your working for this part of the question.

$units = \dfrac{units\ on\ y\text{-axis}}{units\ on\ x\text{-axis}} = \dfrac{cm^3}{min}$

$= 16.7\ cm^3/min$

Q2. This graph shows how the velocity of a car changes over time.
Determine the initial velocity of the car. [1 mark]

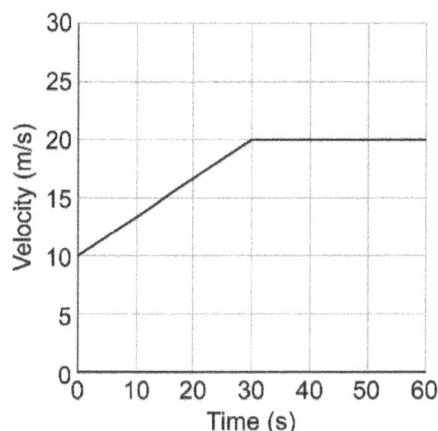

Gradients and intercepts

Identify which axis the question relates to:

The question is looking for the velocity of the car. This is on the y-axis. (The word 'initial' tells us they are looking for the starting velocity.)

Read the scale where the line crosses the axis: 10 m/s [1 mark]

Faded examples

Q1. The graph shows the change in velocity of a car over time.

Determine the acceleration of the car by finding the gradient of the line. [4 marks]

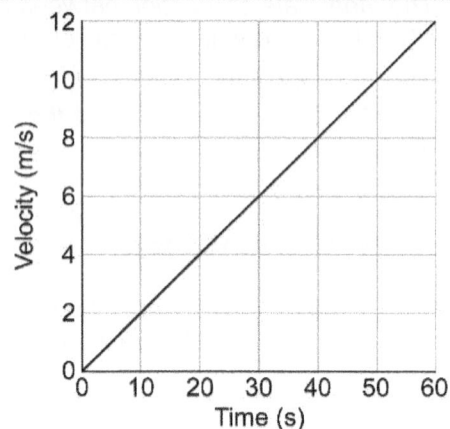

Find values for Δy and Δx: $\quad\Delta y = 12\text{m/s} - 0\text{ m/s} = 12\text{ m/s}$ [1 mark]

$\Delta x = 60\text{ s} - 0\text{ s} + 60\text{ s}$

Substitute values into equation: $\quad$ gradient $= \dfrac{\Delta y}{\Delta x}$ [1 mark]

$\text{gradient} = \dfrac{12}{60}$

Calculate answer: $\quad$ ... [1 mark]

Write units: $\quad$ ... [1 mark]

Q2. A student monitored the reaction between marble chips and dilute hydrochloric acid.

Determine the rate of reaction over the first 20 seconds. [4 marks]

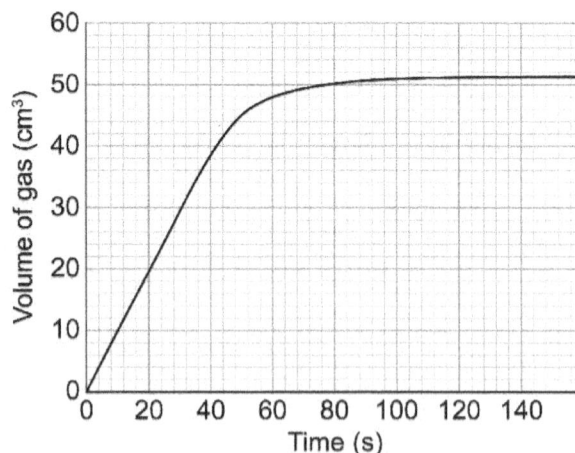

Find values for Δy and Δx: $\quad\Delta y = 20\text{ cm}^3 - 0\text{ cm}^3 = 20\text{ cm}^3$ [1 mark]

$\Delta x = 20\text{ s} - 0\text{ s} = 20\text{ s}$

Substitute values into equation: $\quad$ ... [1 mark]

Calculate answer: $\quad$ ... [1 mark]

Write units: $\quad$ ... [1 mark]

Q3. A student investigated the effect of concentration of salt solution on the mass of potato chips.

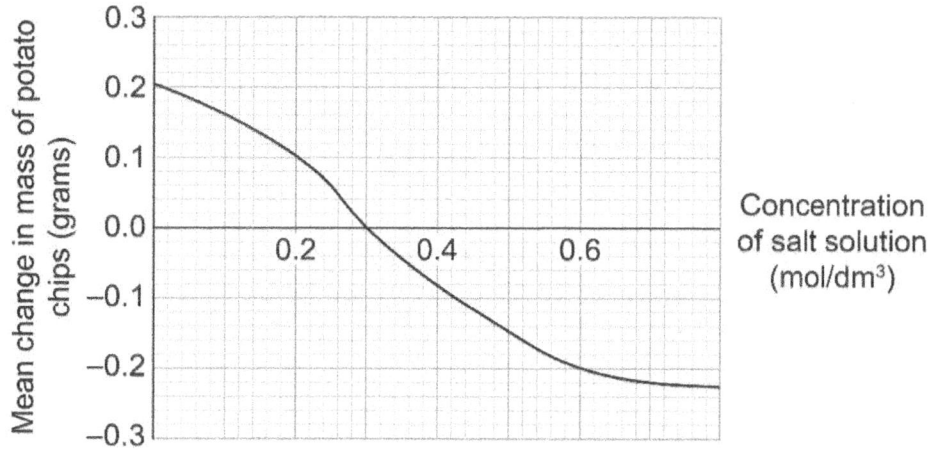

Use the graph to estimate the concentration of the solution inside the potato cells. [1 mark]

Identify which axis the question relates to: The question is looking for the
concentration of solution.
This is on the x-axis.

Read the scale where the line crosses the axis: ...   [1 mark]

Biology questions

Q1. Use the graph to answer the following questions.

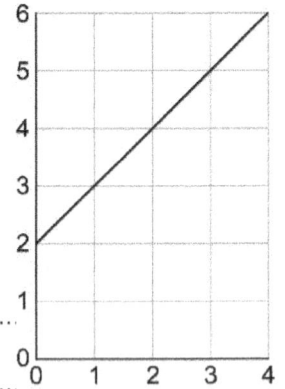

a) Determine the gradient of the line. You do not need to include units.

....

....

....[3 marks]

b) Identify the y-intercept. .... [1 mark]

Q2. A student investigated the activity of catalase on the decomposition of hydrogen peroxide.

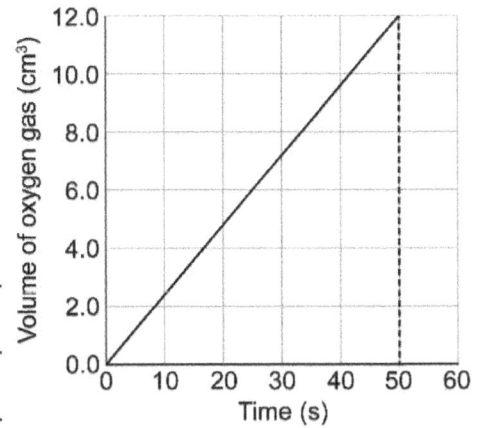

Calculate the rate of reaction using the gradient of the line.

....

....

....

....[4 marks]

Q3. A student investigated the rate of growth of bacterial cells in a nutrient solution at 20 °C. The graph shows the results.

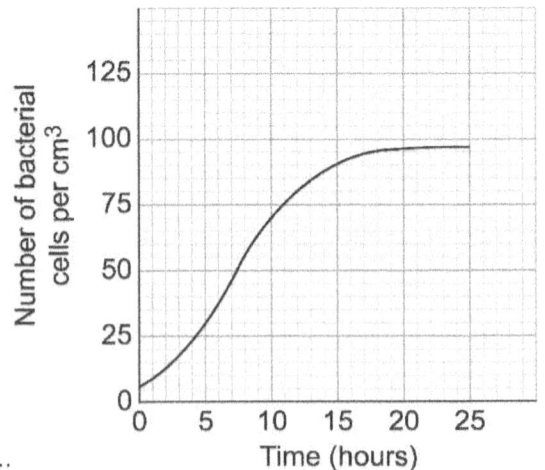

a) Calculate the mean rate of growth of bacterial cells between 5 and 10 hours.
Use the units: number of cells per cm^3 per hour.

....

....

....[3 marks]

b) How many bacterial cells per cm^3 were present at the start of the experiment?

.... [1 mark]

Q4. A student measured the rate of transpiration in the same plant under different conditions by measuring water uptake in a potometer.

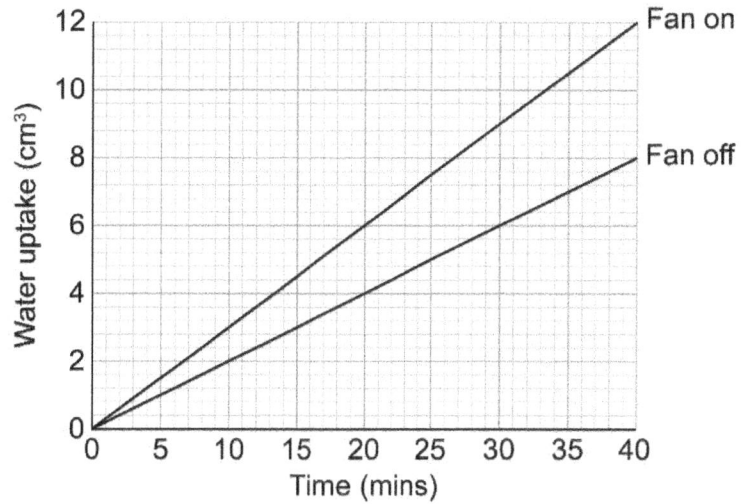

a) Calculate the rate of transpiration when the fan is on.

....

....

....

....

....

....[4 marks]

b) Calculate the rate of transpiration when the fan is off.

....

....

....

....[4 marks]

Q5. A student investigated the activity of an enzyme at different temperatures.
The graph shows the concentration of the product of the reaction at 20 °C.

Calculate the rate of reaction.

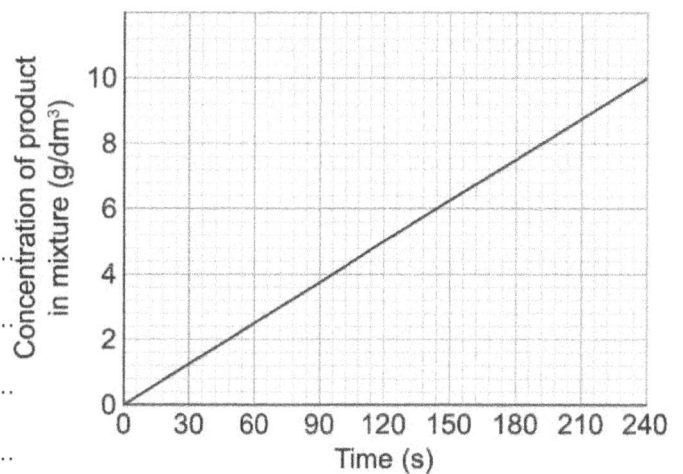

....

....

....

....

....

....[4 marks]

[Total marks 24]

Chemistry questions

Q1. Use the graph to answer Questions **a)** and **b)**.

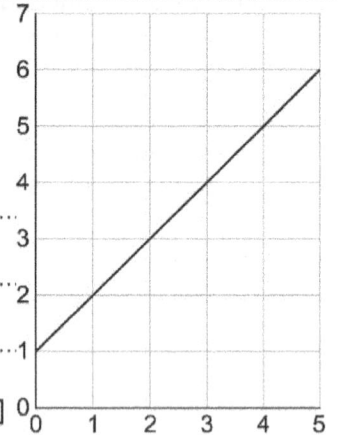

a) Determine the gradient of the line. You do not need to include units.

....

....

....

....[3 marks]

b) Identify the y-intercept. .... [1 mark]

Q2. A student monitored the rate of a chemical reaction over time.

Calculate the rate of reaction using the gradient of the line.

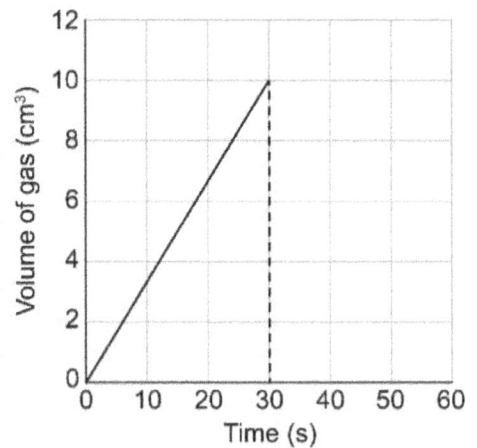

....

....

....

....

....[4 marks]

Q3. A student monitored the rate of a chemical reaction by collecting hydrogen gas over time.

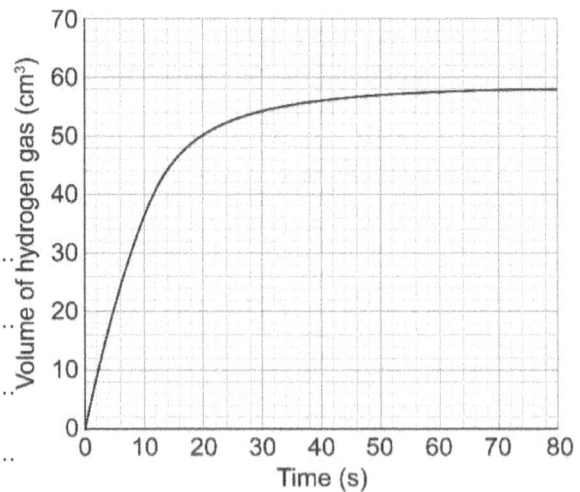

a) Determine the rate of reaction in the first 8 seconds.

....

....

....

....

[4 marks]

b) Determine the mean rate of reaction between 10 seconds and 20 seconds.

....

....

....

....[4 marks]

Q4. A student calculated the number of moles of hydrogen gas produced during a chemical reaction.

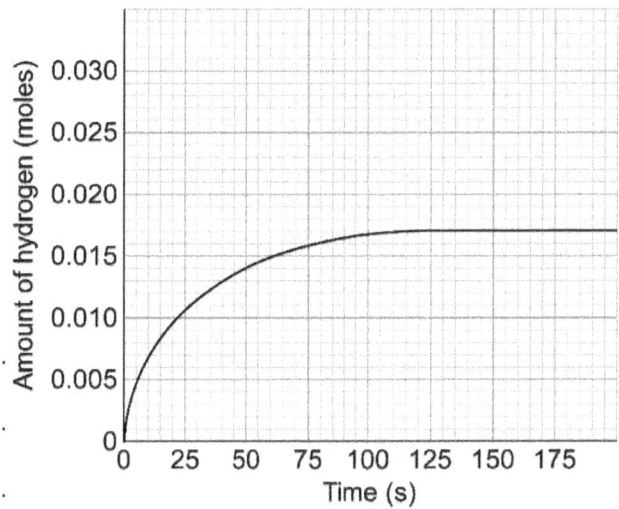

Determine the mean rate of reaction between 25 seconds and 80 seconds.

....

....

....

....

....[4 marks]

Q5. The graph below shows how the volume of a gas changes with temperature.

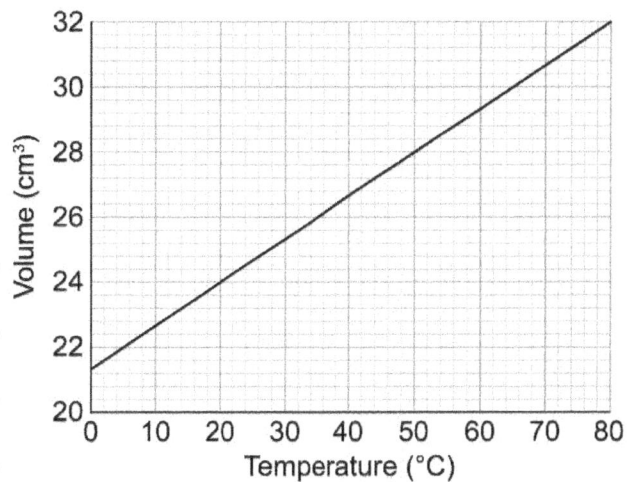

a) Calculate the gradient of the line.

....

....

....

....

....[4 marks]

b) Use the graph to find the volume of the gas at 0 °C.

....[1 mars]

[Total marks 25]

Physics questions

Q1. Use the graph to answer the following questions.

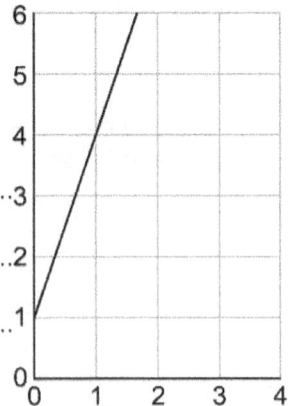

a) Determine the gradient of the line. You do not need to include units.

....

....

....

.... [3 marks]

b) Identify the y-intercept. .... [1 mark]

Q2. A student investigated the motion of a trolley by measuring the distance it travelled at regular time intervals.

Calculate the speed of the trolley using the gradient of the line.

....

....

....

.... [4 marks]

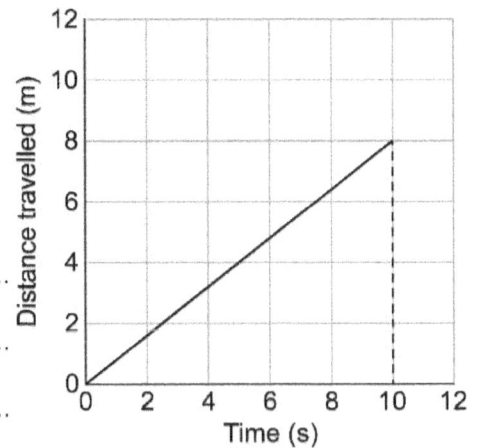

Q3. This is a velocity-time graph for a journey along a straight road.

a) Calculate the acceleration over the first 2 seconds.

....

....

....

.... [4 marks]

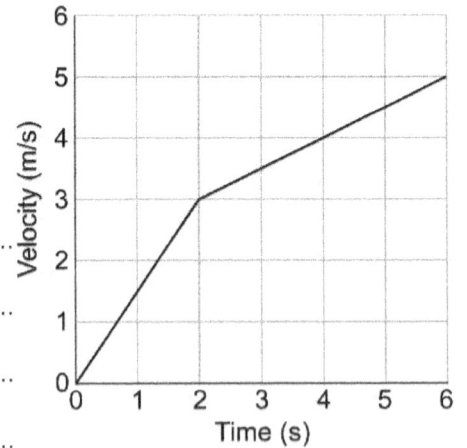

b) Calculate the acceleration over the final 4 seconds.

....

....

.... [4 marks]

Q4. This graph shows the velocity of an object over time.

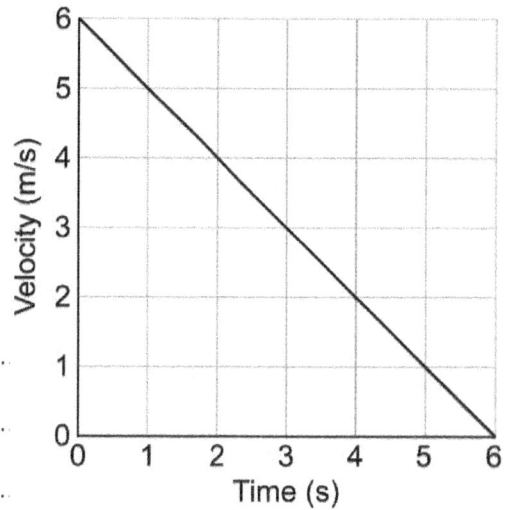

Calculate the deceleration (or negative acceleration) of the object.

....

....

....

....

....[4 marks]

Q5. This graph shows how the velocity of a car changes over time.

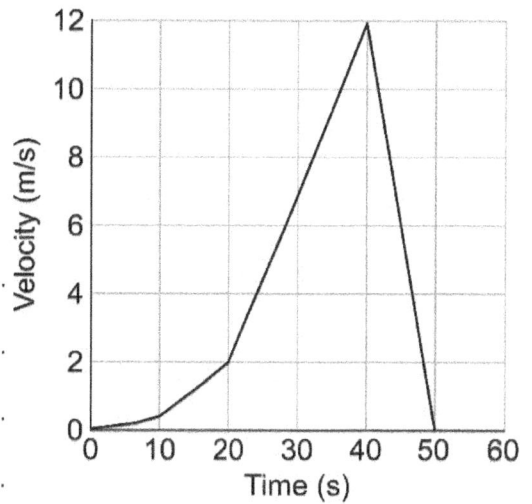

a) Calculate the acceleration of the car between 20 seconds and 40 seconds.

....

....

....

....

....[4 marks]

b) Calculate the deceleration of the car over the final 10 seconds.

....

....

....

....[4 marks]

[Total marks 28]

Tangents HT

On a curved line, the gradient is changing all the time. Tangents can be used to find the gradient at a particular point of a curved line. A tangent is a straight line that just touches the curve at that particular point. The slope of the tangent should match the slope of the curve at that point.

The gradient of the tangent can be found using the equation: gradient = $\frac{\text{change in } y}{\text{change in } x} = \frac{\Delta y}{\Delta x}$

There are no questions in this section for biology or physis.

Worked examples

Q1. A student investigated the rate of reaction between magnesium and dilute hydrochloric acid.
A tangent to the line has been drawn at 25 seconds.

Determine the rate of reaction at 25 seconds. [4 marks]

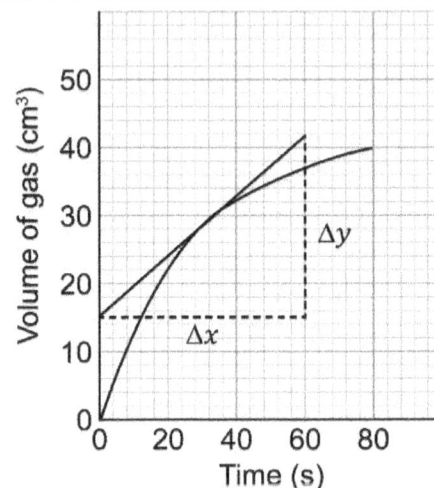

| Use the tangent to find values for Δy and Δx: | Δy = 42 cm³ − 15 cm³ = 27 cm³ [1 mark] |

Use the tangent to find values for Δy and Δx:

Δy = 42 cm^3 − 15 cm^3 = 27 cm^3 [1 mark]

Δx = 60 s − 0 s = 60 s

Substitute values into equation:

gradient = $\frac{\Delta y}{\Delta x}$

gradient = $\frac{27}{60}$ [1 mark]

Calculate answer:

Identify the units using information from the labels on the scales and the equation for the gradient. You do not need to show your working for this part of the question. [1 mark]

Write units:

units = $\frac{\text{units on } y\text{-axis}}{\text{units on } x\text{-axis}} = \frac{cm^3}{min}$

= 0.45 cm^3/s [1 mark]

Q2. A student investigated the rate of a reaction by measuring the decrease in mass over time.

Determine the rate of reaction at 15 seconds.
You should draw a tangent on the graph. [5 marks]

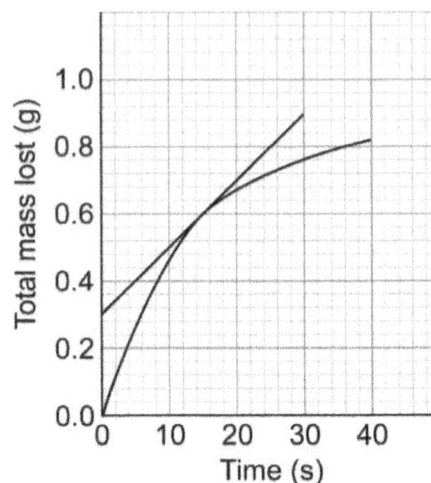

Draw a tangent to the line:	Draw a tangent that touches the curve at 15 s. This has been drawn on the graph. Add a triangle underneath the tangent to help find the values for Δy and Δx.	[1 mark]

Use the tangent to find values for Δy and Δx:

$$\Delta y = 0.90\ g - 0.30\ g = 0.60\ g \qquad \text{[1 mark]}$$

$$\Delta x = 30\ s - 0\ s = 30\ s$$

Substitute values into equation:

$$gradient = \frac{\Delta y}{\Delta x} \qquad \text{[1 mark]}$$

Calculate answer:

$$gradient = \frac{0.06\ g}{30\ s}$$

$$= 0.02 \qquad \text{[1 mark]}$$

Write units:

Identify the units using information from the labels on the scales and the equation for the gradient. You do not need to show your working for this part of the question.

$$units = \frac{\text{units on } y\text{-axis}}{\text{units on } x\text{-axis}} = \frac{g}{s}$$

$$0.02\ g/s \qquad \text{[1 mark]}$$

Faded examples

Q1. A student monitored the reaction between marble chips and dilute hydrochloric acid.

A tangent to the line has been drawn at 52 seconds.

Calculate the rate of reaction at 52 seconds.
[4 marks]

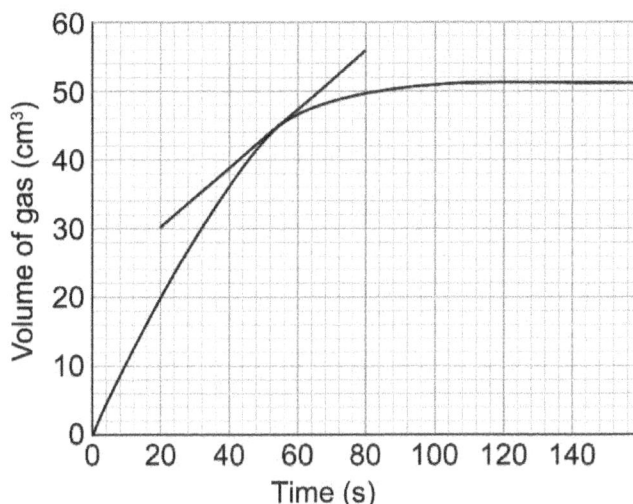

Use the tangent to find values for Δy and Δx:

$$\Delta y = 56\ cm^3 - 30\ cm^3 = 26\ cm^3 \qquad \text{[1 mark]}$$

$$\Delta x = 80\ s - 20\ s = 60\ s$$

Substitute values into equation:

$$gradient = \frac{\Delta y}{\Delta x}$$

$$gradient = \frac{26}{60} \qquad \text{[1 mark]}$$

Calculate answer:

... . [1 mark]

Write units:

$$units = \frac{\text{units on } y\text{-axis}}{\text{units on } x\text{-axis}}$$

... . [1 mark]

Tangents

Q2. A student monitored the rate of a reaction by collecting hydrogen gas over time.

Determine the rate of reaction at 26 seconds.
You should draw a tangent on the graph. [5 marks]

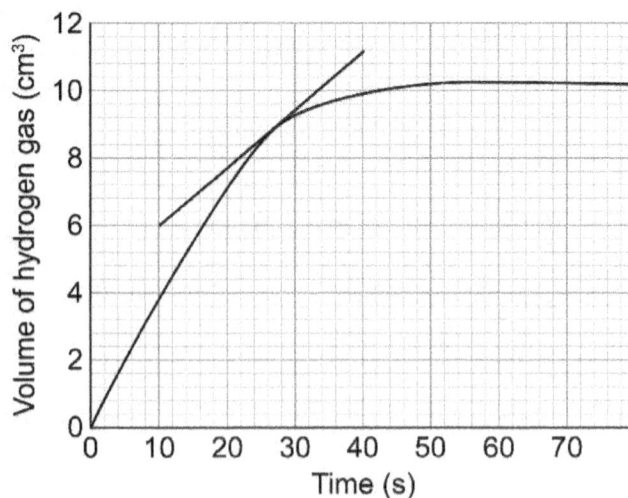

Draw a tangent to the line:

Draw a tangent line that touches the curve at 26 s. This has been drawn on the graph. Add a triangle underneath the tangent to help find the values for Δy and Δx. [1 mark]

Use the tangent to find values for Δy and Δx:

... . [1 mark]

Substitute values into equation:

... . [1 mark]

Calculate answer: ... . [1 mark]

Write units: ... . [1 mark]

Chemistry questions

Q1. A student monitored the reaction between calcium carbonate and hydrochloric acid.
A tangent to the line has been drawn at 22 seconds.

Calculate the rate of reaction at 22 seconds.

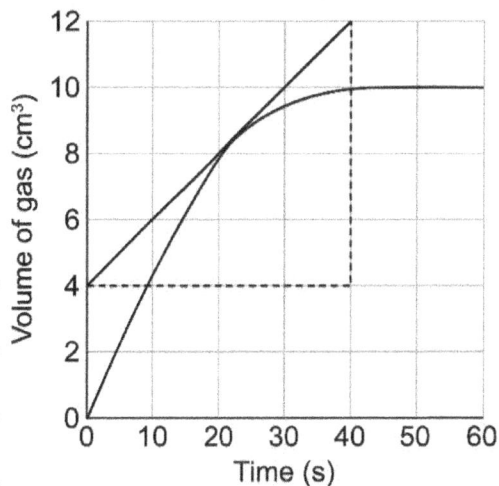

....

....

....

....

....[4 marks]

Q2. A student investigated the rate of a reaction by monitoring the decrease in mass over time.

A tangent to the line has been drawn at 65 seconds.

Calculate the rate of reaction when the time was 65 seconds.

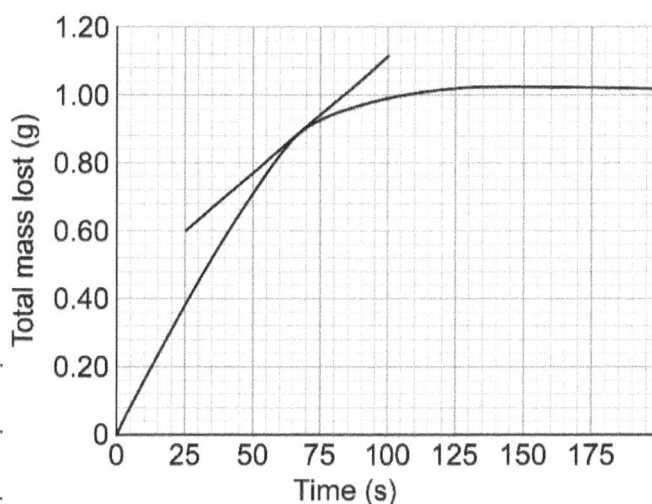

....

....

....

....[4 marks]

Q3. A student monitored the rate of a reaction by collecting hydrogen over time.

A tangent has been drawn on the line of best fit.

Calculate the rate of reaction at this point.

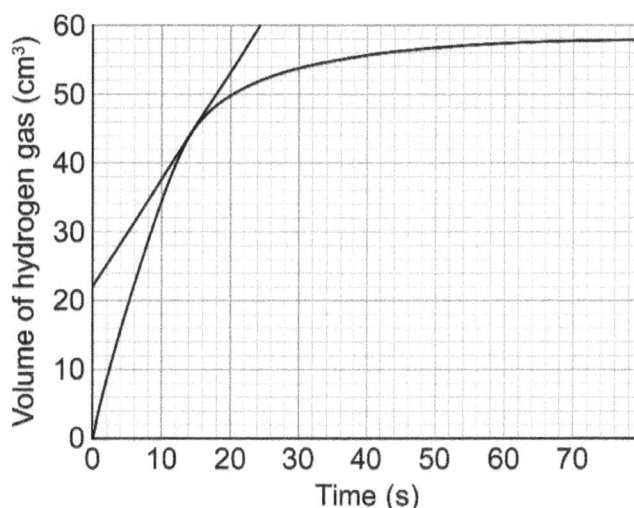

....

....

....

....

....[4 marks]

Tangents

Q4. A student calculated the number of moles of hydrogen gas produced during a chemical reaction.
A tangent to the line has been drawn at 25 seconds.

Calculate the rate of reaction at 25 seconds.

....

....

....

....[4 marks]

Q5. A student investigated the rate of reaction between zinc and dilute sulfuric acid.

Calculate the rate of reaction at 50 seconds.
You should draw a tangent on the graph.

....

....

....

....

....[5 marks]

Q6. A student investigated the reaction between small marble chips and hydrochloric acid.

Calculate the rate of reaction at 40 seconds.
You should draw a tangent on the graph.

....

....

....

....

....[5 marks]

[Total marks 26]

Area under a velocity-time graph HT

The area under a velocity-time graph represents the distance travelled by an object (or displacement of an object). Distance is a scalar quantity; it has magnitude (size) but no direction. Displacement is a vector quantity; it has magnitude and direction.

We can find the area under a velocity-time graph by counting squares or by using the areas of shapes. If the area between the line and the x-axis forms a rectangle then we use the following formula to find the area:
area × length × width*. If the area forms a triangle we use the following formula:* **area =$\frac{1}{2}$× base × height***. Most examination questions can be answered by finding the area of these shapes. If the line is curved then counting squares may be needed:* **area = approximate number of squares × area of one square***.*

As we are finding the distance (or displacement) the units will be for distance e.g. m, not area e.g. m^2.

There are no questions in this section for biology or chemistry.

Worked examples

Q1. This graph shows how the velocity of a car changes over time.

Determine the total distance travelled by the car.   [3 marks]

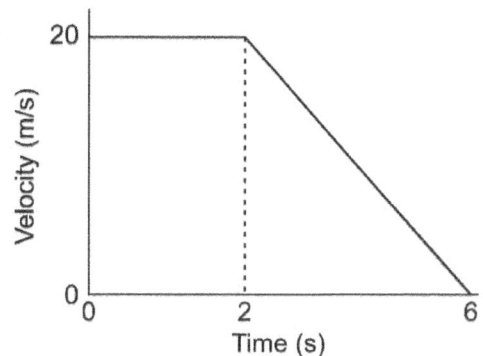

Identify the area and divide into shapes: The question is looking for the total area under the line. This area can be divided into a rectangle and a triangle.

Find the area of the shapes: Distance travelled during first 2 seconds:

$$area = length × width$$

$$area = 2 × 20$$

$$area = 40$$

Distance travelled during the last 4 seconds:

$$area = \frac{1}{2} \text{ base × height}$$

$$area = \frac{1}{2} \text{ } 4 × 20$$

$$area = 40$$ [1 mark]

Find the total area: Total area = 40 + 40 = 80   [1 mark]

Add units: We can see that the unit for velocity on the graph is m/s. This means the distance must be in metres.

80 m   [1 mark]

Area under a velocity–time graph

Q2. A car is studied over a short journey.
A graph of this journey can be seen on the right.

Determine the total distance travelled by the car. [3 marks]

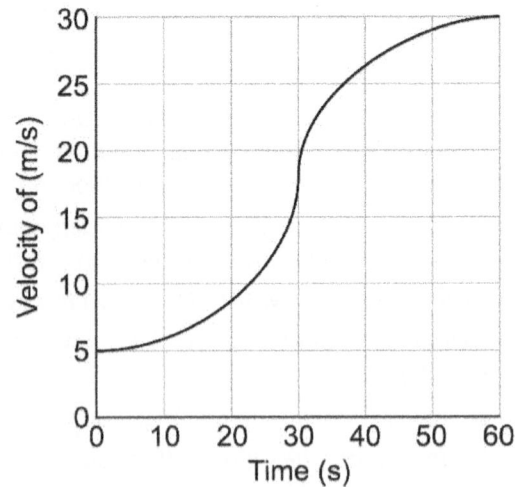

Identify the area and divide into shapes: The question is looking for the total area under the line. The line is curved so we need to count the number of squares.

Find the area of the shapes: Area of one square:

$$area = length \times width$$
$$area = 10 \times 5$$
$$area = 50 \qquad \text{[1 mark]}$$

ind the total area: Estimate the total number of squares:
approximately 20 squares
area = number of squares × area of one square
$$area = 20 \times 50$$
$$area = 1000 \qquad \text{[1 mark]}$$

Add units: We can see that the unit for velocity on the graph is m/s. This means the distance must be in metres.

1000 m [1 mark]

Faded examples

Q1. This graph shows how the velocity of a car changes over time.

Determine the total distance travelled by the car.   [2 marks]

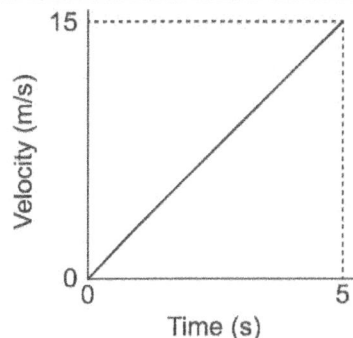

Identify the area and divide into shapes: The question is looking for the total area under the line. This area is a triangle.

Find the area of the shapes: Distance travelled:

$$area = \frac{1}{2} \text{ base} \times \text{height}$$

...

...   .   [1 mark]

Find the total area: ...   .

Add units: ...   .   [1 mark]

Q2. This graph shows how the velocity of a car changes over time.

Determine the total distance travelled by the car.   [3 marks]

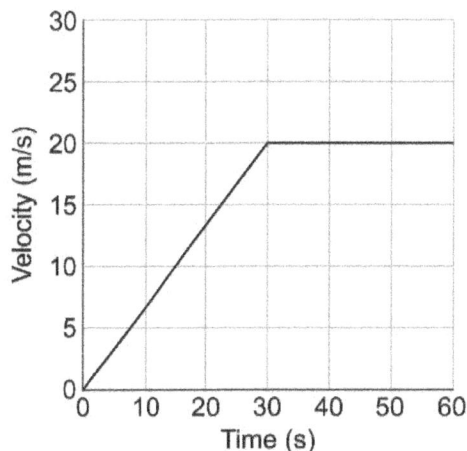

Identify the area and divide into shapes: The question is looking for the total area under the line. This area can be divided into a triangle and a rectangle.

Find the area of the shapes: Distance travelled during first 30 seconds:

...

...

...

Distance travelled during the last 30 seconds:

...

...

...   [1 mark]

Find the total area: ...   .   [1 mark]

Add units: ...   [1 mark]

Area under a velocity-time graph

Q3. This graph shows how the velocity of a skier changes over time.

Determine the total distance travelled by the skier. [3 marks]

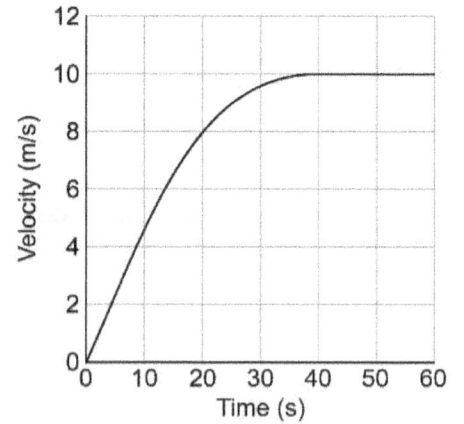

Identify the area and divide into shapes: The question is looking for the total area under the line. The line is curved so we need to count the number of squares.

Find the area of the shapes: area of one square:

area = length × width

…

Estimate the total number of squares:

…

area = number of squares × area of one square

… [1 mark]

Find the total area: … . [1 mark]

Add units: … [1 mark]

Q4. This graph shows how the velocity of an object changes over time.

Determine the total distance travelled by the object. [3 marks]

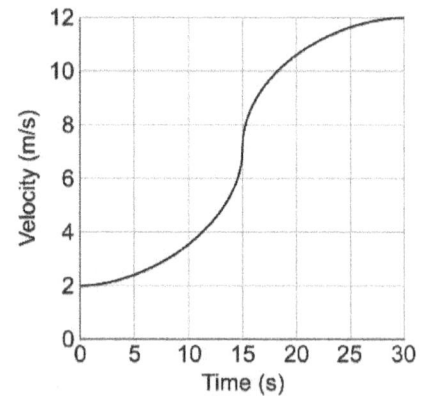

Identify the area and divide into shapes: The question is looking for the total area under the line. The line is curved so we need to count the number of squares.

Find the area of the shapes: area of one square:

…

…

Estimate the total number of squares:

… [1 mark]

Find the total area: … . [1 mark]

Add units: … [1 mark]

Physics questions

Q1. This graph shows the velocity of an object over time.
The object travels at a steady velocity of 10 m/s.

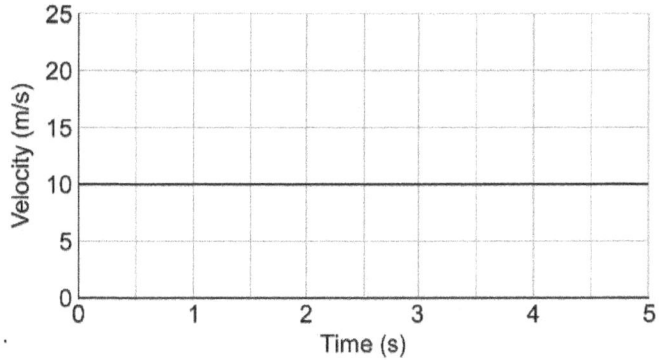

Determine the total distance travelled by the object in 5 seconds.

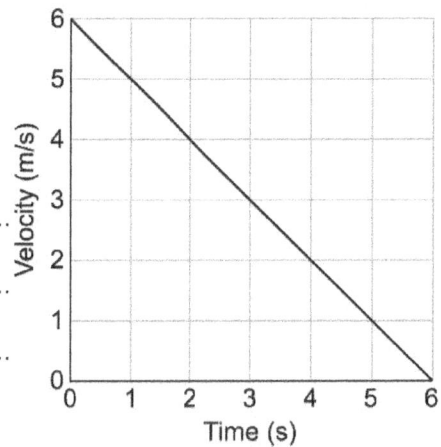

....

....

....[2 marks]

Q2. This graph shows the velocity of an object over time.

Determine the total distance travelled by the object.

....

....

....

....[2 marks]

Q3. This graph shows the velocity of a car over time.

a) Determine the distance travelled in the first 3 seconds.

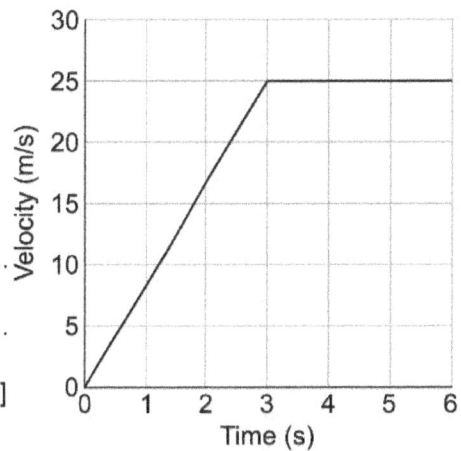

....

....

....[2 marks]

b) Determine the distance travelled in the last 3 seconds.

....

....[2 marks]

Area under a velocity-time graph

Q4. This graph shows how the velocity of a car changes.
The car travels at 10 m/s for 2 seconds and then begins to slow down.

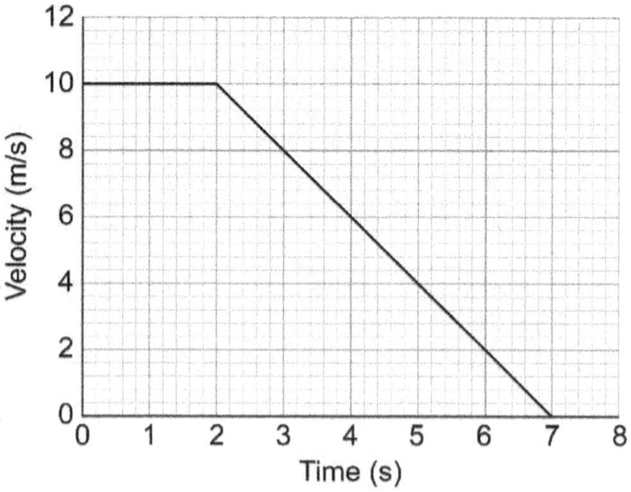

Determine the total distance travelled by the car.

....

....

....

....

....[3 marks]

Q5. This graph shows a journey.

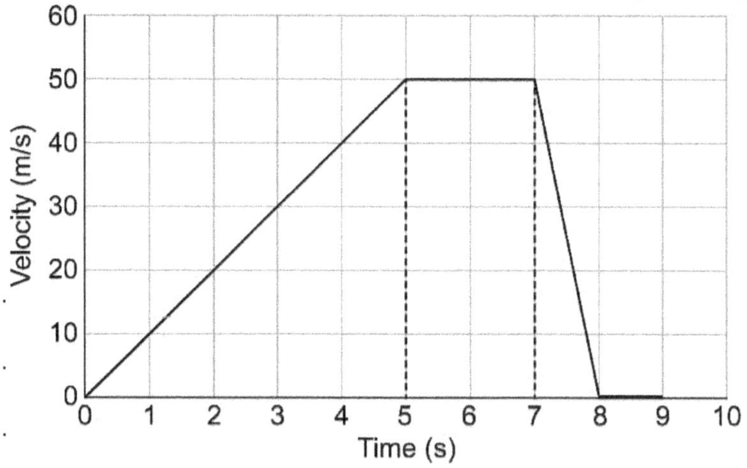

Determine the total distance travelled during the journey.

....

....

....

....[3 marks]

Q6. This graph shows how the velocity of a car changes over time.

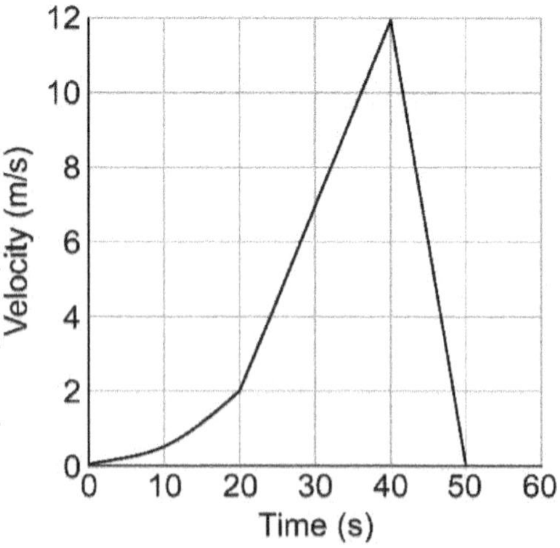

Determine the distance the car travels while decelerating.

....

....

....

.... [2 marks]

Q7. This graph shows the journey of a car.

Determine the total distance travelled by the car.

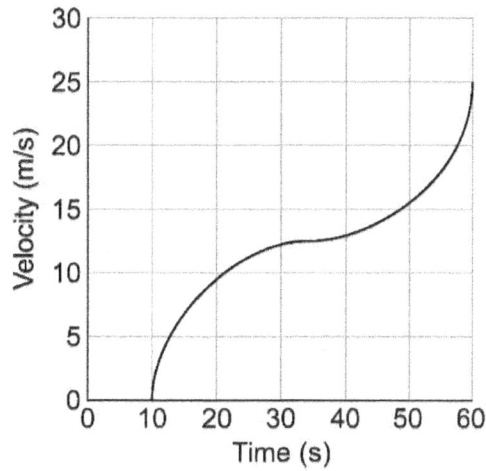

....

....

.... [3 marks]

Q8. This graph shows the journey of a train.

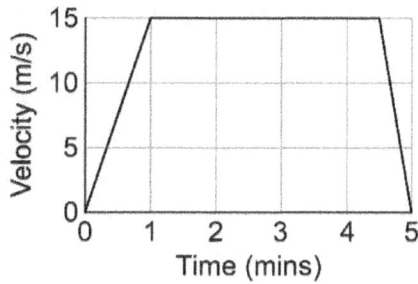

Determine the total distance travelled whilst the train travelled at a steady velocity for 3.5 minutes.
You will need to convert minutes into seconds. There are 60 seconds in 1 minute.

....

....

....

....[3 marks]

[Total marks 22]

Chapter 5 Geometry and trigonometry

Using angles

*Measuring angles is important in physics for understanding light reflection, light refraction, projectile motion and vector diagrams (**HT**). When measuring angles in light reflection and refraction, the dashed line is called the 'normal' and is drawn perpendicular (at 90° or at right angles) to the boundary. To measure the angle of incidence, angle of reflection or angle of refraction, measure the angle between the light ray and the normal line. The magnitude of a resultant force is its size, measured in newtons. When producing scaled vector diagrams, it is useful to use either blocked paper or graph paper.*

There are no questions in this section for biology or chemistry.

Worked examples

Q1. A student investigated the reflection of light by a mirror using a ray box.
Measure the angle of incidence. [1 mark]

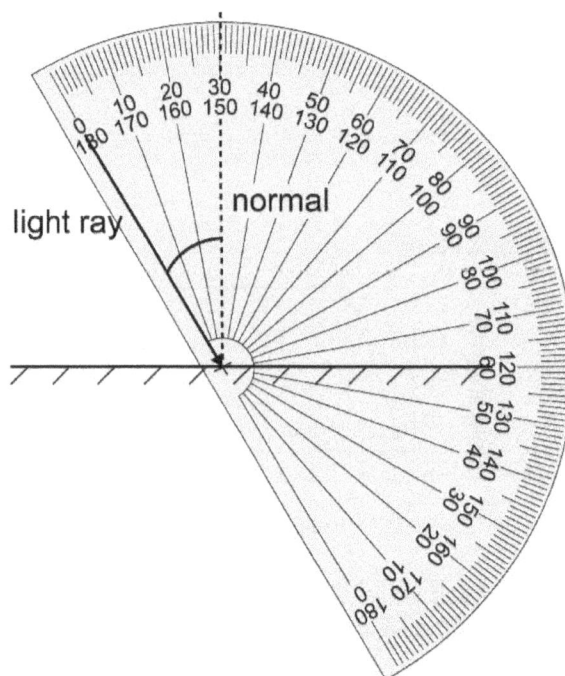

Identify the angle being measured.
Place the cross where the ray and normal meet.
Line up the zero line on the protractor to the ray of light:

The angle of incidence is between the ray of light and the normal. (See diagram)

Read the scale from the ray of light at zero around to the normal: 30° [1 mark]

Q2. **HT** A resultant force has a horizontal component of 40 N and a vertical component of 30 N. Determine the magnitude and direction of the resultant force by drawing a vector diagram. [4 marks]

Draw the horizontal and vertical forces drawn to the same scale
e.g. 1 cm = 10 N.

Add an arrow to form a triangle:

[2 marks]

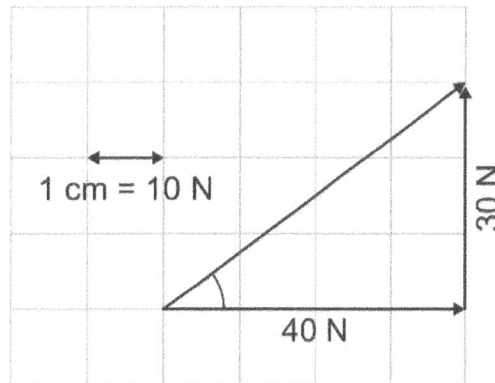

1 cm = 10 N

30 N

40 N

For the magnitude, measure the length of the line in cm and convert to N.

For the direction, measure the angle.

The arrow is 5 cm long.

The magnitude of the resultant force is 50 N. [1 mark]

The angle is 37°

The direction of the resultant force is 37° from the horizontal force. [1 mark]

Faded example

Q1. **HT** A resultant force has a horizontal component of 30 N and a vertical component of 10 N. Determine the magnitude and direction of the resultant force by drawing a vector diagram. [3 marks]

Draw the horizontal and vertical forces drawn to the same scale e.g.
1 cm = 10 N.

Add an arrow to form a triangle:

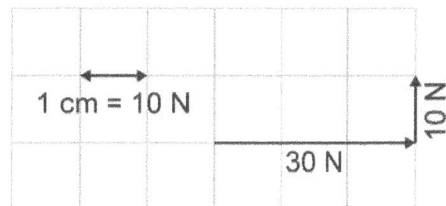

1 cm = 10 N

30 N

10 N

[1 mark]

For the magnitude, measure the length of the line in cm and convert to N. ... . [1 mark]

For the direction, measure the angle. ... . [1 mark]

Physics questions

Q1. A student investigated the refraction of light at the boundary between air and glass.

    a) Label the angle of incidence using the letter *i* on the diagram.   [1 mark]

    b) Label the angle of refraction using the letter *r* on the diagram.   [1 mark]

    c) Measure the angle of incidence.

    ....[1 mark]

    d) Measure the angle of refraction.

    ....[1 mark]

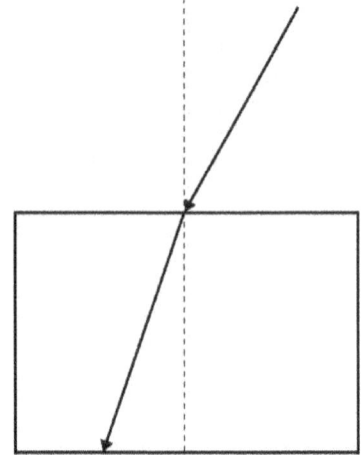

Q2. **HT** A resultant force has a horizontal component of 60 N and a vertical component of 80 N. Determine the magnitude and direction of the resultant force.   [4 marks]

Q3. **HT** A crane applies two forces: 300 N horizontally to the right and 450 N vertically upwards. Determine the magnitude and direction of the resultant force by drawing a vector diagram.   [4 marks]

Q4. **HT** A rescue helicopter winch cable exerts a force with components of 850 N horizontally and 1200 N vertically upwards. Determine the magnitude and direction of the resultant force by drawing a vector diagram.   [4 marks]

[Total marks 16]

Area and surface area to volume ratio

In science you need to know how to calculate the area of triangles and rectangles. This may be applied during field work when we need to know the size of the area being sampled, when calculating pressure or measuring the area under a curve. It is important to be able to calculate the surface area to volume ratio of different reacting solids or gas exchange surfaces in living organisms. Cubes are often used to represent organisms or cells.

$$\text{area of a triangle} = \tfrac{1}{2} \text{ base} \times \text{height} \qquad \text{area of a rectangle} = \text{height} \times \text{width}$$

$$\text{surface area of a cube} = 6 \times \text{area of one face} \textbf{ or } (\text{length of side})^2 \times 6$$

$$\text{volume of a cube (or cuboid)} = \text{length} \times \text{height} \times \text{width}$$

Worked examples

Q1. A triangle has a base of 6 cm and a height of 4 cm.
Calculate the area of the triangle. [3 marks]

Write down equation:	area of a triangle $= \tfrac{1}{2}$ base × height	
Substitute values into equation:	area of a triangle $= \tfrac{1}{2}$ 6 × 4	[1 mark]
Calculate answer:	area of a triangle $= \tfrac{1}{2} \times 24 = 12$	[1 mark]
Write units:	12 cm^2	[1 mark]

Q2. **a)** A cube has a height of 2 cm. Calculate the surface area of the cube. [3 marks]

Write down equation:	surface area of a cube = 6 × area of one face	
Calculate the area of one face of the cube:	area of one face = 2 × 2 = 4	[1 mark]
Calculate answer:	surface area of the cube = 4 × 6 = 24	[1 mark]
Write units:	24 cm^2	[1 mark]

b) Calculate the volume of the cube. [3 marks]

Write down equation:	volume of a cube = length × height × width	
Substitute values into the equation:	volume of the cube = 2 × 2 × 2	[1 mark]
Calculate answer:	volume of the cube = 8	[1 mark]
Write units:	8 cm^2	[1 mark]

c) Find the surface area to volume ratio for the cube. [1 mark]

Write down the values:	surface area : volume = 24 : 8	
Substitute values into the equation:	surface area : volume = 3 : 1	[1 mark]

Faded examples

Q1. A triangular sail has a base of 9 m and a height of 7 m.
Calculate the area of the sail that catches the wind. [2 marks]

Write down equation: area of a triangle = $\frac{1}{2}$ base × height

Substitute values into equation: area of a triangle = $\frac{1}{2}$ 9 × 7

Calculate answer: area of a triangle = $\frac{1}{2}$ 63 = ... [1 mark]

Write units: ... [1 mark]

Q2. A wooden block has a length of 20 cm and a width of 15 cm.
Calculate the area of contact when the block is placed flat on a table. [2 marks]

Write down the relevant equation: area of a rectangle = height × width

Substitute values into equation: ...

Calculate answer: ... [1 mark]

Write units: ... [1 mark]

Biology questions

Q1. A student is carrying out field work to sample the different types of plant species using a quadrat. The quadrat is 25 cm long and 25 cm wide.

Calculate the area of one quadrat. ....

....[2 marks]

Q2. A model bacterial cell is a cube with sides of length 4.0 µm.

a) Calculate the surface area of the model bacterial cell. ....

....[3 marks]

b) Calculate the volume of the model bacterial cell. ....

....[3 marks]

c) Calculate the surface area to volume ratio. ....

.... [1 mark]

Q3. In an investigation into enzyme activity, a student cuts a potato into cubes of different sizes.
Cube **A** has sides which measure 1.5 cm.

a) Calculate the surface area of cube **A**. ....

....[3 marks]

b) Calculate the volume of cube **A** ....

....[3 marks]

c) Calculate the surface area to volume ratio of cube **A**. ....

.... [1 mark]

Cube **B** has sides which measure 3.0 cm.

d) Calculate the surface area of cube **B**. ....

....[3 marks]

e) Calculate the volume of cube **B**. ....

....[3 marks]

f) Calculate the surface area to volume ratio of cube **B**. ....

.... [1 mark]

Q4. A student is investigating the distribution of dandelions in a field.
Calculate the total area of the field.

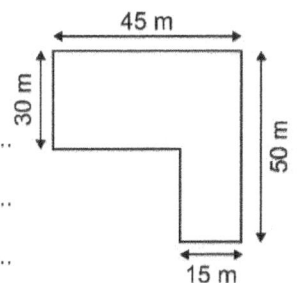

....

....

....

....[4 marks]

[Total marks 27]

Chemistry questions

Q1. A student investigated the effect of surface area on the rate of the reaction between calcium carbonate and dilute hydrochloric acid. A cube of calcium carbonate has a side length of 50 mm.

a) Calculate the total surface area of the cube of calcium carbonate...

....

....[3 marks]

b) Calculate the volume of the cube of calcium carbonate. ....

....

....[3 marks]

c) Calculate the surface area to volume ratio of the cube.
Give your answer as the simplest whole number ratio. ....

.... [1 mark]

Q2. A cubic nanoparticle has a side length of 2 nm.

a) Calculate the total surface area of the nanoparticle..

....

....[3 marks]

b) Calculate the volume of the nanoparticle. ....

....[3 marks]

c) Calculate the surface area to volume ratio of the nanoparticle.
Give your answer as the simplest whole number ratio. ....

....

.... [1 mark]

Q3. A cubic nanoparticle has a side length of 1.5 nm.

a) Calculate the total surface area of the nanoparticle..

....

....[3 marks]

b) Calculate the volume of the nanoparticle. ....

....

....[3 marks]

c) Calculate the surface area to volume ratio of the nanoparticle.
Give your answer as the simplest whole number ratio. ....

.... [1 mark]

[Total marks 21]

Physics questions

Q1. A cube has a side length of 5 cm.

    a) Calculate the total surface area of the cube...

    ....[3 marks]

    b) Calculate the volume of the cube. ....

    ....[3 marks]

Q2. A student calculated the pressure exerted by a cube of wood on a table: pressure $= \dfrac{\text{force}}{\text{area}}$

Part of the calculation involves finding the area of the base of the cube.
Each side of the cube has a length of 50 mm.
Calculate the area of the surface of the cube in contact with the table. ....

    ....[2 marks]

Q3. A student calculated the density of a cube of wood: density $= \dfrac{\text{mass}}{\text{volume}}$

Part of the calculation involves finding the volume of the cube.
Each side of the cube has a length of 25 mm.
Calculate the volume of the cube. ....

    ....

    ....[3 marks]

Q4. A student investigated the heat transfer through a cuboid material.
Part of the experiment involves finding the surface area of the cuboid.
Calculate the surface area of the cuboid.

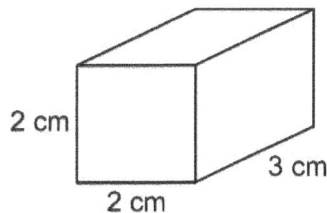

2 cm   3 cm   2 cm

    ....

    ....

    ....[3 marks]

Q5. A solar panel has a triangular section with a base of 8.4 m and a height of 6.5 m.
Calculate the area of the panel exposed to sunlight.

    ....

    ....

    ....[3 marks]

[Total marks 17]

Arithmetic

Standard form

Mark scheme for faded examples

Q1. Divide the numbers and subtract the powers:

$\dfrac{3 \times 10^8}{3 \times 10^9} = 1 \times 10^{8-9}$ [1]

Write the answer: $= 1 \times 10^{-1}$ m [1]

Q2. Substitute variables into equation:

magnification $= \dfrac{2 \times 10^{-4}}{2 \times 10^{-5}}$ [1]

Divide the numbers and subtract the powers:

$\dfrac{2 \times 10^{-4}}{2 \times 10^{-5}} = 1 \times 10^{-4-(-5)}$ [1]

Write the answer: $= 1 \times 10^1 = \times 10$ [1]

Mark scheme for biology questions

Q1. 5×10^6 [1]

Q2. 8.6×10^{10} [1]

Q3. 5×10^{-6} [1]

Q4. 8.5×10^{-7} [1]

Q5. 0.0000005 [1]

Q6. magnification $= \dfrac{4 \times 10^{-4}}{2 \times 10^{-5}}$ [1]

$\dfrac{4 \times 10^{-4}}{2 \times 10^{-5}} = 2 \times 10^{-4-(-5)}$ [1]

$= 2 \times 10^1$ or $\times 20$ [1]

Q7. $(3.0 \times 10^{13}) \times (2.0 \times 10^4)$ [1]

$= 6.0 \times 10^{13+4}$ [1]

$= 6.0 \times 10^{17}$ [1]

Q8. $(2.0 \times 10^5) \times (1.0 \times 10^2)$ [1]

$= 2.0 \times 10^{5+2}$ [1]

$= 2.0 \times 10^7$ [1]

Q9. $\dfrac{4.2 \times 10^{-5}}{1.4 \times 10^{-6}}$ [1]

$= 3.0 \times 10^{-5-(-6)}$ [1]

$= 3.0 \times 10^1$ or 30 times larger [1]

Mark scheme for chemistry questions

Q1. 7.2×10^2 [1]

Q2. 4×10^4 [1]

Q3. 1×10^{-3} [1]

Q4. 2.5×10^{-2} [1]

Q5. 9.8×10^{-8} [1]

Q6. concentration $= \dfrac{2.5 \times 10^{-2}}{5.0 \times 10^{-2}}$ [1]

$\dfrac{2.5 \times 10^{-2}}{5.0 \times 10^{-2}} = 0.5 \times 10^{-2-(-2)}$ [1]

$= 0.5 \times 10^0$ or 0.5 [1]

Q7. concentration $= \dfrac{0.1}{2.5 \times 10^{-2}}$ or $\dfrac{1.0 \times 10^{-1}}{2.5 \times 10^{-2}}$ [1]

$\dfrac{1.0 \times 10^{-1}}{2.5 \times 10^{-2}} = 0.4 \times 10^{-1-(-2)}$ [1]

$= 0.4 \times 10^1$ or 4 [1]

Q8. mass of substance $= 1 \times 10^{-3} \times 23$ [1]

$1 \times 10^{-3} \times 23 = 23 \times 10^{-3}$ [1]

$= 2.3 \times 10^{-2}$ g [1]

Q9. mass of substance $= 5.5 \times 10^{-3} \times 11$ [1]

$5.5 \times 10^{-3} \times 11 = 60.5 \times 10^{-3}$ [1]

$= 6.05 \times 10^{-2}$ or 6.1×10^{-2} g [1]

Mark scheme for physics questions

Q1. 1×10^8 [1]

Q2. 1.1×10^8 [1]

Q3. 5×10^{-7} [1]

Q4. 1×10^{-10} [1]

Q5. wavelength $= \dfrac{3.0 \times 10^8}{5.5 \times 10^{14}}$ [1]

$\dfrac{3.0 \times 10^8}{5.5 \times 10^{14}} = 0.545 \times 10^{8-14} = 0.545 \times 10^{-6}$ [1]

$= 5.5 \times 10^{-7}$ m (accept 5.45×10^{-7}) [1]

Q6. wavelength $= \dfrac{3.0 \times 10^8}{2.4 \times 10^9}$ [1]

$\dfrac{3.0 \times 10^8}{2.4 \times 10^9} = 1.25 \times 10^{8-9}$ [1]

$= 1.25 \times 10^{-1}$ m (accept 0.125 or 1.3×10^{-1}) [1]

Q7. wavelength $= \dfrac{3.0 \times 10^8}{3.0 \times 10^{-7}}$ [1]

$\dfrac{3.0 \times 10^8}{3.0 \times 10^{-7}} = 1 \times 10^{8-(-7)}$ [1]

$= 1 \times 10^{15}$ m [1]

Ratios

Mark scheme for faded examples

Q1. Simplify: $3 : 2$ [1]

Q2. Substitute values into ratio: $5 : 15$

Simplify: $1 : 3$ [1]

Q3. Calculate the phenotypic ratio: $1 : 0$ [1]

Q4. Write down the phenotype of each offspring:

black; black; white; white

Calculate the phenotypic ratio: 1 : 1 [1]

Mark scheme for biology questions

Q1. 600 : 75 which is simplified to 8 : 1 [1]

Q2. A = 6 : 1; B = 3 : 1; C = 3 : 2 or 1.5 : 1 [3]

Q3. 1 : 1 [1]

Q4. 12 : 6 simplified to 2 : 1 [1]

Q5. 6 : 1 [1]

Q6. a) 1 : 1 [1]

b) 1 : 0 [1]

c) 3 : 1 [1]

Mark scheme for chemistry questions

Q1. 1 : 2 [1]

Q2. 2 : 3 [1]

Q3. 1 : 1 [1]

Q4. 1 : 2 [1]

Q5. 65 : 35 simplified to 13 : 7 [1]

Q6. a) 1 : 2 [1]

b) 3 : 7 [1]

c) 8 : 17 [1]

d) 1 : 2 [1]

Q7. 24 : 8 simplified to 3 : 1 [1]

Q8. 1.5 : 0.125 simplified to 12 : 1 [1]

Q9. $NaCl$ [1]

Mark scheme for physics questions

Q1. 86 : 136 simplified to 43 : 68 [1]

Q2. 36 : 12 simplified to 3 : 1 [1]

Q3. 1200 : 400 simplified to 3 : 1 [1]

Q4. 1.2 : 6 simplified to 1 : 5 [1]

Q5. 85 : 15 simplified to 17 : 3 [1]

Q6. 300 : 50 simplified to 6 : 1 [1]

Q7. 32 : 30 simplified to 16 : 15 [1]

Q8. 25 : 1 [1]

Q9. 63 : 27 simplified to 7 : 3 [1]

Q10. The potential difference ratio is the same as the resistance ratio 12 : 18 : 6

(accept 8 V : 12 V : 4 V) [1]

Simplified ratio 2 : 3 : 1 [1]

Percentages and percentage change

Mark scheme for faded examples

Q1. Multiply by 100 to find the percentage change:

0.06 × 100 = 6% [1]

Q2. Divide the difference by the original:

$\frac{0.1}{2.0}$ = 0.05 [1]

Multiply by 100 to find the percentage change:

0.05 × 100 = 5% [1]

Mark scheme for biology questions

Q1. $\frac{30}{120}$ × 100 = 25% [1]

Q2. $\frac{15}{20}$ × 100 = 75% [1]

Q3. $\frac{50}{50\,000}$ × 100 = 0.1% [1]

Q4. $\frac{1.02 - 1.2}{1.2}$ × 100 [1]

= −15% [1]

Q5. $\frac{0.97 - 0.85}{0.85}$ × 100 [1]

= 14% [1]

Q6. Number of bacteria killed

= 250 000 − 175 000 = 75 000 [1]

$\frac{75\,000}{250\,000}$ × 100 = 30% [1]

Q7. $\frac{550 - 400}{400}$ × 100 [1]

= 37.5% [1]

Mark scheme for chemistry questions

Q1. 12 + 16 = 28 [1]

$\frac{12}{28}$ × 100 = 43% (accept 42.9%) [1]

Q2. 14 + 1 + 1 + 1 = 17 [1]

$\frac{14}{17}$ × 100 = 82% (accept 82.4%) [1]

Q3. (2 × 14) + (4 × 1) + (3 × 16) = 80 [1]

$\frac{28}{80}$ × 100 = 35% [1]

Q4. a) % yield = $\frac{500}{1500}$ × 100 [1]

= 33.3% [1]

b) % yield = $\frac{750}{1500}$ × 100 [1]

= 50% [1]

c) % yield = $\frac{1000}{1500}$ × 100 [1]

= 66.7% [1]

d) % yield = $\frac{1200}{1500}$ × 100 [1]

= 80% [1]

Q5. atom economy = $\frac{112}{244}$ × 100 [1]

= 46% (accept 45.9%) [1]

Answers

Q6. atom economy = $\frac{106}{217} \times 100$ [1]

    = 49% (accept 48.8%)   [1]

Mark scheme for physics questions

Q1. 100 − 15 = 85%   [1]

Q2. $\frac{4}{8} \times 100 = 50\%$ [1]

Q3. a) $\frac{84.1 - 82.3}{82.3} \times 100$ [1]

    = 2.19%   [1]

   b) $\frac{14.8 - 13.9}{13.9} \times 100$ [1]

    = 6.47%   [1]

   c) $\frac{40.6 - 40.6}{40.6} \times 100$ [1]

    = 0%   [1]

   d) total for 2024 = 84.1 + 14.8 + 40.6 = 139.5

    total for 2023 = 82.3 + 13.9 + 40.6 = 136.8

    $\frac{139.5 - 136.8}{136.8} \times 100$ [1]

    = 1.97%   [1]

Q4. $\frac{240 - 12}{240} \times 100$ [1]

    = 95%   [1]

Q5. $\frac{85}{100} \times 200$ [1]

    = 170 J   [1]

Q6. percentage efficiency = $\frac{70}{200} \times 100$ [1]

    = 35%   [1]

Q7. percentage efficiency = $\frac{500}{850} \times 100$ [1]

    = 59% (accept 58.8%)   [1]

Q8. percentage efficiency = $\frac{264}{1200} \times 100 = 22.0\%$ [1]

    difference = 25.0% − 22.0% = 3.0%   [1]

Rates

Mark scheme for faded examples

Q1. Calculate answer and give units:

    $\frac{280}{2}$ = 140 beats per minute   [1]

Q2. Substitute values into equation:

    average rate = $\frac{22 \text{ breaths}}{2 \text{ minutes}}$ [1]

    Calculate answer and give units:

    $\frac{22}{2}$ = 11 breaths per minute   [1]

Q3. Calculate answer and give units:

    $\frac{6}{40}$ = 0.15 g/s   [1]

Q4. Substitute values into equation:

    average rate = $\frac{50 - 40 \text{ cm}^3}{80 - 40 \text{ s}}$ [1]

    Calculate answer and give units:

    $\frac{10}{40}$ = 0.25 cm³/s   [1]

Mark scheme for biology questions

Q1. heart rate = $\frac{525}{7}$ [1]

    = 75 beats per minute   [1]

Q2. transpiration rate = $\frac{6}{30}$ [1]

    = 0.2 cm³/min   [1]

Q3. a) heart rate = $\frac{650}{5}$ [1]

    = 130 beats per minute   [1]

   b) heart rate = $\frac{720}{10}$ [1]

    = 72 beats per minute   [1]

Q4. a) rate of reaction = $\frac{8}{2}$ [1]

    = 4 cm³/min   [1]

   b) rate of reaction = $\frac{15}{5}$ [1]

    = 3 cm³/min   [1]

Q5. a) rate of oxygen bubble production = $\frac{115}{5}$ [1]

    = 23 bubbles/min   [1]

   b) rate of oxygen bubble production = $\frac{55}{5}$ [1]

    = 11 bubbles/min   [1]

Q6. rate of water uptake = $\frac{68}{9}$ [1]

    = 7.6 mm/min   [1]

Q7. rate of blood flow in A = $\frac{15}{10}$ [1]

    = 1.5 cm³/s   [1]

    rate of blood flow in B = $\frac{12}{15}$ [1]

    = 0.8 cm³/s   [1]

Mark scheme for chemistry questions

Q1. rate of reaction = $\frac{16 \text{ cm}^3}{20 \text{ s}}$ [1]

    = 0.8 cm³/s   [1]

Q2. rate of reaction = $\frac{70 \text{ cm}^3}{100 \text{ s}}$ [1]

    = 0.7 cm³/s   [1]

Q3. rate of reaction = $\frac{0.008 \text{ mol}}{16 \text{ s}}$ [1]

    = 0.0005 mol/s (accept 5×10^{-4} mol/s)   [1]

Q4. rate of reaction = $\frac{152 \text{ g} - 151 \text{ g}}{200 \text{ s}}$ [1]

    = 0.05 g/s (accept 5.0×10^{-2})   [1]

Q5. a) rate of reaction $= \frac{16 \text{ cm}^3}{20 \text{ s}}$ [1]

$= 0.8 \text{ cm}^3\text{/s}$ [1]

b) rate of reaction $= \frac{48 - 16 \text{ cm}^3}{100 - 20 \text{ s}}$ [1]

$= 0.4 \text{ cm}^3\text{/s}$ [1]

Q6. a) rate of reaction $= \frac{0.12 \text{ g}}{10 \text{ min}}$ [1]

$= 0.012 \text{ g/min}$ [1]

b) rate of reaction $= \frac{0.24 - 0.12 \text{ g}}{20 - 10 \text{ min}}$ [1]

$= 0.012 \text{ g/min}$ [1]

Handling data

Significant figures

Mark scheme for faded examples

Q1. Calculate the answer: 0.94 A [1]

Q2. Round up or round down:

Round down as decider digit is less than 5

Calculate the answer: 750 000 m [1]

Mark scheme for biology questions

Q1. a) 9760 and 9800 [1]

b) 3730 and 3700 [1]

c) 0.266 and 0.27 [1]

d) 3.71 and 3.7 [1]

e) 0.00373 and 0.0037 [1]

Q2. 1680 daisy plants [1]

Q3. 46 stomata [1]

Q4. a) 21 900 m^2 [1]

b) 22 000 m^2 [1]

Q5. a) 0.163 s [1]

b) 0.16 s [1]

Q6. 37.4 °C [1]

Q7. a) 186 000 red blood cells [1]

b) 190 000 red blood cells [1]

Q8. 2.85×10^3 bacteria [1]

Mark scheme for chemistry questions

Q1. a) 5880 and 5900 [1]

b) 4740 and 4700 [1]

c) 0.346 and 0.35 [1]

d) 6.51 and 6.5 [1]

e) 0.00233 and 0.0023 [1]

Q2. 70% [1]

Q3. 58.4 g/dm^3 [1]

Q4. a) 3.33 g [1]

b) 3.3 g [1]

Q5. a) 1.67 g [1]

b) 1.7 g [1]

Q6. 0.263 mol/dm^3 [1]

Q7. a) 0.0460 dm^3 [1]

b) 0.046 dm^3 [1]

Q8. 1.7×10^{-27} kg [1]

Mark scheme for physics questions

Q1. a) 7350 and 7400 [1]

b) 5830 and 5800 [1]

c) 0.464 and 0.46 [1]

d) 6.31 and 6.3 [1]

e) 0.00763 and 0.0076 [1]

Q2. 18.4 J [1]

Q3. 1.67 m/s^2 [1]

Q4. a) 8.21 Ω [1]

b) 8.2 Ω [1]

Q5. a) 8.86 s [1]

b) 8.9 s [1]

Q6. 131 W [1]

Q7. a) 0.998 g/cm^3 [1]

b) 1.0 g/cm^3 [1]

Q8. 1.40×10^{10} years [1]

Arithmetic mean

Mark scheme for faded examples

Q1. a) Divide by the number of values to find the mean:

$\frac{60.2}{5} = 12.0$ cm [1]

b) Calculate the sum of the height increases:

$15.1 + 14.8 + 15.5 + 14.9 + 14.7 = 75.0$ [1]

Divide by the number of values to find the mean:

$\frac{75.0}{5} = 15.0$ cm [1]

Q2. Divide by the number of values to find the mean:

$\frac{67.80}{3} = 22.60$ cm^3 [1]

Q3. Find the sum of the concordant data:

$32.65 + 32.75 + 32.70 = 98.10$ [1]

Divide by the number of values to find the mean:

$\frac{98.10}{3} = 32.70$ cm^3 [1]

Answers

Mark scheme for biology questions

Q1. $6.2 + 5.8 + 6.5 + 6.3 = 24.8$ [1]

$\frac{24.8}{4} = 6.2$ cm [1]

Q2. a) $72 + 70 + 74 = 216$ [1]

$\frac{216}{3} = 72$ bpm [1]

b) $126 + 128 + 130 = 384$ [1]

$\frac{384}{3} = 128$ bpm [1]

Q3. $0.42 + 0.39 + 0.41 + 0.40 + 0.35 = 1.97$ [1]

$\frac{1.97}{5} = 0.39$ g (not 0.394 g) [1]

Q4.

For 10 cm: $27 + 28 + 26 = 81$ [1]

$\frac{81}{3} = 27$ bubbles [1]

For 30 cm: $12 + 14 + 13 = 39$ [1]

$\frac{39}{3} = 13$ bubbles [1]

For 50 cm: $3 + 1 + 5 = 9$ [1]

$\frac{9}{3} = 3$ bubbles [1]

Q5. a) sum of values = 626.5 [1]

$\frac{626.5}{8} = 78.3$ g/L [1]

b) The anomaly is 124.8.

sum of values excluding anomaly = 501.7 [1]

$\frac{501.7}{7} = 71.7$ g/L (accept 71.67) [1]

Mark scheme for chemistry questions

Q1. $44 + 42 + 43 = 129$ [1]

$\frac{129}{3} = 43$ cm^3 [1]

Q2. $17.6 + 17.2 + 17.8 + 17.4 = 70.0$ [1]

$\frac{70.0}{4} = 17.5$ °C [1]

Q3. $0.63 + 0.62 + 0.66 + 0.65 = 2.56$ [1]

$\frac{2.56}{4} = 0.64$ g [1]

Q4. $2.59 + 2.58 + 2.60 + 2.58 + 2.55 = 12.9$ [1]

$\frac{12.9}{5} = 2.58$ g [1]

Q5. concordant data: 24.45; 24.55; 24.50

$24.45 + 24.55 + 24.50 = 73.50$ [1]

$\frac{73.50}{3} = 24.50$ cm^3

(do not accept 24.425 or 24.43) [1]

Mark scheme for physics questions

Q1. $15.2 + 15.8 + 15.5 + 15.4 = 61.9$ [1]

$\frac{61.9}{4} = 15.5$ s [1]

Q2. $2.0 + 1.9 + 1.7 + 1.8 = 7.4$ [1]

$\frac{7.4}{4} = 1.9$ A (accept 1.85 A) [1]

Q3. $0.45 + 0.42 + 0.46 + 0.43 + 0.44 = 2.2$ [1]

$\frac{2.2}{5} = 0.44$ m/s [1]

Q4. For A: $84 + 82 + 85 + 83 = 334$ [1]

$\frac{334}{4} = 84\%$ (accept 83.5%) [1]

For B: $76 + 74 + 75 = 225$ [1]

$\frac{225}{3} = 75\%$ [1]

For C: $68 + 70 + 67 + 69 = 274$ [1]

$\frac{274}{4} = 69\%$ (accept 68.5%) [1]

Sampling

Mark scheme for faded examples

Q1. Substitute values into equation:

population estimate = 2 × 2400 [1]

Calculate the answer:

2 × 2400 = 4800 daisy plants [1]

Q2. Calculate the total area sampled:

20 × 0.25 = 5 m^2 [1]

Calculate the density:

$\frac{45}{5} = 9$ [1]

Write the answer:

9 orchids per m^2 [1]

Mark scheme for biology questions

Q1. area of quadrat = 1 × 1 = 1 m^2 [1]

total area of field = 100 × 80 = 8000 m^2 [1]

total area sampled = 20 × 1 = 20 m^2 [1]

density = $\frac{85}{20} = 4.25$ organisms per m^2 [1]

population estimate = 4.25 × 8000 = 34 000 [1]

Q2. area of quadrat = 1 × 1 = 1 m^2 [1]

total area of field = 50 × 30 = 1500 m^2 [1]

total area sampled = 20 × 1 = 20 m^2 [1]

density = $\frac{97}{20} = 4.85$ organisms per m^2 [1]

population estimate = 4.85 × 1500 = 7275 [1]

Q3. area of quadrat = 0.5 × 0.5 = 0.25 m^2 [1]

total area of field = 75 × 45 = 3375 m^2 [1]

total area sampled = 32 × 0.25 = 8 m^2 [1]

density = $\frac{96}{8} = 12$ organisms per m^2 [1]

population estimate = 12 × 3375 = 40 500 [1]

Q4. area of quadrat = 0.5 × 0.5 = 0.25 m² [1]

total area sampled = 12 × 0.25 = 3 m² [1]

density = $\frac{36}{3}$ [1]

12 buttercups per m² [1]

Q5. a) area of quadrat = 0.5 × 0.5 = 0.25 m² [1]

b) total area sampled = 10 × 0.25 = 2.5 m² [1]

c) density of A = $\frac{25}{2.5}$ = 10 snails per m² [1]

density of B = $\frac{30}{2.5}$ = 12 snails per m² [1]

d) population estimate A = 10 × 472 = 4720 [1]

population estimate B = 12 × 720 = 8640 [1]

Simple probability

Mark scheme for faded examples

Q1. Convert into the answer:

The probability of offspring having yellow fruit is ½

(0.5, 50% , 2 : 1 or 1 in 2). [1]

Q2. Substitute variables into the equation: $\frac{1}{4}$

Convert into the answer:

The probability of polled offspring is 0.75 [1]

Mark scheme for biology questions

Q1. accept $\frac{1}{2}$, 0.5, 50%, 2 : 1 or 1 in 2 [1]

Q2. 25% [1]

Q3. Punnett square completed correctly (All Bb) [1]

	B	B
b	Bb	Bb
b	Bb	Bb

100% [1]

Q4. Punnett square completed correctly. [1]

	R	r
r	Rr	rr
r	Rr	rr

0.5 [1]

Q5. 100% [1]

Q6. Punnett square completed correctly. [1]

	R	r
R	RR	Rr
r	Rr	rr

accept $\frac{3}{4}$, 0.75, 75%, 3 : 1 or 3 in 4 [1]

Q7. a) Genetic diagram completed correctly. [1]

Parent alleles

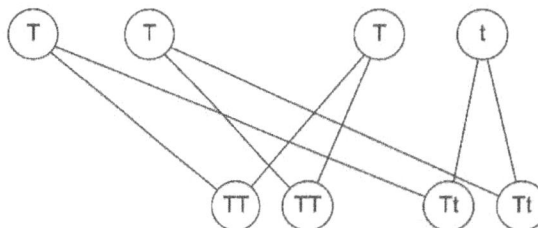

Offspring genotypes	TT	TT	Tt	Tt
Offspring phenotypes	Tall	Tall	Tall	Tall

b) 0% [1]

Q8. 25% [1]

Q9. $\frac{1}{2}$ [1]

Q10. $\frac{4}{6}$ simplified to $\frac{2}{3}$ [1]

Q11. a) 12 : 8 simplified to 3 : 2 [1]

b) $\frac{12}{20}$ simplified to $\frac{3}{5}$ [1]

Converting units

Mark scheme for faded examples

Q1. Calculate the answer including units:

1500 ÷ 1000 = 1.5 kg [1]

Q2. Is the unit getting larger or smaller?

Larger. This means we divide by the conversion factor.

Calculate the answer including units:

1.5 mg ÷ 1000 = 0.0015 g [1]

Q3. Calculate the answer including units:

125 ÷ 1000 = 0.125 dm³ [1]

Q4. Is the unit getting larger or smaller?

Smaller. This means we multiply by the conversion factor.

Calculate the answer including units:

0.5 × 1000 = 500 cm³ [1]

Answers

Mark scheme for biology questions

Q1. 7.5 µm × 1000 = 7500 nm [1]

Q2. 2000 nm ÷ 1000 = 2 µm [1]

Q3. a) 4300 nm [1]

    b) 27 000 nm [1]

    c) 375 nm [1]

    d) 120 000 nm [1]

Q4. a) 0.125 mm [1]

    b) 0.0075 mm [1]

    c) 0.0005 mm [1]

    d) 0.059 mm [1]

Q5. a) 0.2 µm [1]

    b) 0.0002 mm [1]

    c) 200 ÷ 100 000 = 0.00002 cm [1]

Q6. a) 120 µm [1]

    b) 120 000 nm [1]

Q7. 2500 nm [1]

Q8. 6.2 dm^3 × 1000 = 6200 cm^3 [1]

Q9. 0.25 m^2 × 10 000 = 2500 cm^2 [1]

Q10. 850 cm^2 ÷ 10 000 = 0.085 m^2 [1]

Mark scheme for chemistry questions

Q1. 55 nm ÷ 1000 = 0.055 µm [1]

Q2. 25.30 ÷ 1000 = 0.0253 dm^3 [1]

Q3. a) 2000 g [1]

    b) 40 000 g [1]

    c) 3200 g [1]

    d) 350 g [1]

Q4. a) 6 kJ [1]

    b) 5.2 kJ [1]

    c) 0.0103 kJ [1]

    d) 0.0005 kJ [1]

Q5. a) 5000 mg [1]

    b) 0.0004 g [1]

    c) 0.1 g [1]

    d) 10 mg [1]

Q6. a) 0.0296 dm^3 [1]

    b) 200 cm^3 [1]

Q7. 0.025 dm^3 [1]

Mark scheme for physics questions

Q1. 250 ÷ 100 = 2.5 m [1]

Q2. 1.37 m × 100 = 137 cm [1]

Q3. a) 2.5 A [1]

    b) 1.050 A [1]

    c) 0.73 A [1]

    d) 15 A [1]

Q4. a) 5000 Ω [1]

    b) 79 000 Ω [1]

    c) 2600 Ω [1]

    d) 500 Ω [1]

Q5. a) 250 000 µm [1]

    b) 0.0009 g [1]

    c) 0.102 km [1]

    d) 21 800 m [1]

Q6. a) 0.378 dm^3 [1]

    b) 730 cm^3 [1]

    c) 1000 000 Hz [1]

    d) 2000 000 000 000 W [1]

Algebra

Solving equations

Mark scheme for faded examples

Q1. Simplify the equation: $\frac{\text{charge flow}}{\text{current}}$ = time [1]

Q2. Multiply both sides of the equation by volume:

density × volume = $\frac{\text{mass}}{\text{volume}}$ × volume

Write answer: density × volume = mass [1]

Q3. Calculate answer: $\frac{60}{12}$ = 5 [1]

Write units: 5 A [1]

Q4. Change the subject of the equation (if needed):

$\frac{490}{5 \times 9.8} = \frac{5 \times 9.8 \times h}{5 \times 9.8}$

$\frac{490}{5 \times 9.8} = h$ [1]

Calculate the answer: $\frac{490}{49}$ = 10 [1]

Write units: 10 m [1]

Q5. Calculate answer: $\frac{32\,000}{0.5 \times 10^3}$ = 640 [1]

Write units: 640 kg [1]

Q6. Change the subject of the equation (if needed):

$\frac{7900}{0.5 \times 5^2} = \frac{0.5 \times m \times 10^2}{0.5 \times 10^2}$

$\frac{7900}{0.5 \times 10^2} = m$ [1]

Calculate the answer: $\frac{7900}{0.5 \times 10^2}$ = 632 [1]

Write units: 632 kg [1]

Answers

Mark scheme for biology questions

Q1. magnification × size of real object = size of image [1]

Q2. $\dfrac{\text{total magnification}}{\text{objective lens magnification}}$ = eyepiece lens magnification [1]

Q3. magnification = $\dfrac{2}{0.02}$ [1]

$\dfrac{2}{0.02}$ = 100 [1]

×100 [1]

Q4. magnification = $\dfrac{15\,000}{30}$ [1]

$\dfrac{15\,000}{30}$ = 500 [1]

×500 [1]

Q5. size of real object = $\dfrac{\text{size of image}}{\text{magnification}}$ [1]

$\dfrac{8}{400}$ [1]

0.02 mm [1]

Q6. estimated population = $\dfrac{8}{1}$ × 200 [1]

1600 [1]

dandelions [1]

Q7. size of real object = $\dfrac{\text{size of image}}{\text{magnification}}$ [1]

$\dfrac{20}{400}$ [1]

0.05 mm [1]

Q8. estimated population = $\dfrac{6}{0.5}$ × 150 [1]

1800 [1]

clover plants [1]

Mark scheme for chemistry questions

Q1. R_f × distance moved = distance moved
by solvent   by substance [1]

Q2. concentration × volume = mass [1]

Q3. $R_f = \dfrac{3.4}{5.0}$ [1]

$\dfrac{3.4}{5.0}$ = 0.68 [1]

(no unit, do not accept cm) [1]

Q4. concentration = $\dfrac{2.5}{50}$ [1]

0.05 [1]

g/cm^3 [1]

Q5. 0.5 = $\dfrac{\text{distance moved by substance}}{12.4 \text{ cm}}$ [1]

0.5 × 12.4 = 6.2 [1]

cm [1]

Q6. 0.27 g/cm^3 = $\dfrac{\text{mass}}{250 \text{ cm}^3}$ [1]

0.27 × 250 = 67.5 [1]

g [1]

Q7. 0.05 = $\dfrac{\text{mass of substance}}{12}$ [1]

0.05 × 12 = 0.6 [1]

g [1]

Q8. 0.35 = $\dfrac{\text{mass of substance}}{18}$ [1]

0.35 × 18 = 6.3 [1]

g [1]

Mark scheme for physics questions

Q1. $\dfrac{\text{resultant force}}{\text{acceleration}}$ = mass [1]

Q2. speed = $\dfrac{\text{distance travelled}}{\text{time}}$ [1]

Q3. period = $\dfrac{1}{50}$ [1]

0.02 [1]

s [1]

Q4. 200 = force × 8 [1]

$\dfrac{200}{8}$ = 25 [1]

N [1]

Q5. 3000 = $\dfrac{30\,000}{\text{time}}$ [1]

$\dfrac{30\,000}{3000}$ = 10 [1]

s [1]

Q6. 12 = 3 × R [1]

$\dfrac{12}{3}$ = 4 [1]

Ω [1]

Q7. 0.75 = $\dfrac{150}{\text{total power input}}$ [1]

$\dfrac{150}{0.75}$ = 200 [1]

W [1]

Q8. (final velocity)^2 − (0)^2 = 2 × 2 × 28 [1]

final velocity = $\sqrt{0^2 + (2 \times 2 \times 28)}$

final velocity = $\sqrt{112}$ = 10.6 [1]

m/s [1]

Graphs

Interpreting line graphs

Mark scheme for faded examples

Q1. a) Read value from axis: 30 s [1]

b) Read value from axis: 10 m/s [1]

Answers

Q2. a) Read value from axis: 20 cm^3 [1]

    b) Find the value on the appropriate axis:

    42 cm^3 is found on the y-axis

    Draw lines on graph:

    Arrows or lines drawn on graph

    Read value from axis: 48 s [1]

Q3. Read value using the line of best fit:

    7.6 g/dm^3 [1]

Q4. Read value using the line of best fit: 6 cm^3 [1]

Mark scheme for biology questions

Q1. a) Draw a line from 2 minutes vertically until it reaches the line of best fit. Draw a line to the y-axis and read off the value: 15 bubbles [1]

    b) Read the value from the x-axis when the line of best fit becomes horizontal: 6 minutes [1]

Q2. a) Draw a line from the x-axis at 30 °C and read off value on y-axis: 24 cm^3 [1]

    b) Draw a line from the x-axis at 60 °C and read off value on y-axis: 10 cm^3 [1]

    c) As temperature increases the rate of respiration in yeast increases up to a maximum at 44 °C. [1] The rate then rapidly decreases due to enzymes being denatured in the yeast cells. [1]

Q3. a) Draw a line vertically up from pH 3.0 to the line of best fit and read off the value from the y-axis: 22 minutes [1]

    b) Find the shortest time taken value on the y-axis and draw a horizontal line to the line of best fit and then read off the pH value on the x-axis: pH 2.1 [1]

    c) The enzyme works fastest at pH 2.0. [1]

    At pH values lower than 2.0 or higher than 2.0, digestion takes longer. [1]

Q4. a) Draw a line up to line of best fit from 30 seconds to the 1.0% line of best fit and read off the value on the y-axis: 47 cm^3 [1]

    b) Draw a line horizontally from 40 cm^3 on the y-axis to the line of best fit for 0.5% and draw a vertical line down until it hits the x-axis. Read off the value: 48 s [1]

Q5. When prey numbers are high, predators have plenty of food and their population starts to increase. [1]
As predator numbers increase, they consume more prey causing the prey population to decline. [1]
With fewer prey available, the predator numbers begin to decline. This allows the prey population to recover and the cycle begins again. [1]

Mark scheme for chemistry questions

Q1. a) 8 cm^3 [1]

    b) 40 s [1]

Q2. a) 26 cm^3 [1]

    b) 60 s [1]

    c) The rate of reaction is fastest at the start. [1]

    After approximately 20 seconds the rate of reaction decreases. [1]

    The reaction stops after approximately 80 seconds. [1]

Q3. a) 6 cm^3 (accept between 5 and 7 cm^3) [1]

    b) As temperature increases the rate of reaction increases/ As the temperature increases the volume of gas collected after 30 seconds increases. [1]

    At lower temperature/ between 20 and 30 °C, the rate of increase is lower. [1]

    At higher temperatures; 50–60 °C; the rate of increase is higher. [1]

Q4. a) 10% [1]

    b) At lower pressures as pressure increases, the percentage yield increases. [1]

    At higher pressures the rate of increase in percentage yield is lower. [1]

    c) The higher the temperature the lower the percentage yield of ammonia/ the lower the temperature the higher the percentage yield of ammonia. [1]

Mark scheme for physics questions

Q1. a) Draw a line vertically from the x-axis to the line of best fit and then read off the value from the y-axis: 7 m/s [1]

    b) Read off the value on the y-axis: 20 m/s [1]

    c) Draw a line horizontally from the y-axis to the line of best fit and then read off the value from the x-axis: 15 s [1]

    d) The relationship shows that velocity increases at a constant rate (constant acceleration) [1]

    until it reaches a maximum velocity at 30 seconds.[1] After 30 seconds the velocity remains constant. [1]

Q2. a) Draw a line vertically from the x-axis to the line of best fit and then read off the value from the y-axis: 6 m/s [1]

    b) Find where the line of best fit crosses the x-axis: 7 s [1]

c) The object is travelling at a constant velocity of 10 m/s (no acceleration) between 0 and 2 s. [1]

It then decelerates at a constant rate from 2 seconds onwards [1]

until it reaches 0 m/s after 7 s. [1]

Q3. a) 15 mm [1]

b) 5 mm [1]

c) This shows a linear relationship. [1]

The extension of the spring is directly proportional to the amount of force applied to it. [1]

The larger the applied force the larger the extension of the spring. [1]

Q4. a) 1.6 A [1]

b) 12 V [1]

c) The relationship between current and potential difference for the 5 Ω resistor is linear and directly proportional. [1]

As potential difference is increased, current increases at a constant rate. [1]

d) 1.4 A [1]

Gradients and intercepts

Mark scheme for faded examples

Q1. Calculate answer:

$\frac{12}{60} = 0.2$ [1]

Write units:

units $= \frac{\text{units on } y\text{-axis}}{\text{units on } x\text{-axis}} = \frac{\text{m/s}}{\text{s}} = \frac{\text{m}}{\text{s} \times \text{s}}$

m/s^2 [1]

Q2. Substitute values into equation:

gradient $= \frac{\Delta y}{\Delta x}$

gradient $= \frac{20}{20}$ [1]

Calculate answer:

$\frac{20}{20} = 1$ [1]

Write units:

units $= \frac{\text{units on } y\text{-axis}}{\text{units on } x\text{-axis}} = \frac{\text{cm}^3}{\text{s}}$

$= 1$ cm^3/s [1]

Q3. Read the scale where the line crosses the axis:

0.3 mol/dm^3 [1]

Mark scheme for biology questions

Q1. a) $\Delta y = 6 - 2 = 4$

$\Delta x = 4 - 0 = 4$ [1]

gradient $= \frac{\Delta y}{\Delta x}$

gradient $= \frac{4}{4}$ [1]

$\frac{4}{4} = 1$ [1]

b) 2 [1]

Q2. $\Delta y = 12.0$ cm^3 $- 0.0$ cm^3 $= 12.0$ cm^3

$\Delta x = 50$ s $- 0$ s $= 50$ s [1]

gradient $= \frac{\Delta y}{\Delta x}$

gradient $= \frac{12}{50}$ [1]

$\frac{12}{50} = 0.24$ [1]

units $= \frac{\text{units on } y\text{-axis}}{\text{units on } x\text{-axis}} = \frac{\text{cm}^3}{\text{s}}$

cm^3/s [1]

Q3. a) $\Delta y = 70 - 30 = 40$ number of cells per cm^3

$\Delta x = 10$ hr $- 5$ hr $= 5$ hr [1]

gradient $= \frac{\Delta y}{\Delta x}$

gradient $= \frac{40}{5}$ [1]

$\frac{40}{5} = 8$ [1]

b) 5 bacterial cells/cm^3 (accept 4.8–5.2) [1]

Q4. a) $\Delta y = 12$ cm^3 $- 0.0$ cm^3 $= 12$ cm^3

$\Delta x = 40$ min $- 0$ min $= 40$ min [1]

gradient $= \frac{\Delta y}{\Delta x}$

gradient $= \frac{12}{40}$ [1]

$\frac{12}{40} = 0.3$ [1]

units $= \frac{\text{units on } y\text{-axis}}{\text{units on } x\text{-axis}} = \frac{\text{cm}^3}{\text{min}}$

cm^3/min [1]

b) $\Delta y = 8$ cm^3 $- 0.0$ cm^3 $= 8$ cm^3

$\Delta x = 40$ min $- 0$ min $= 40$ min [1]

gradient $= \frac{\Delta y}{\Delta x}$

gradient $= \frac{8}{40}$ [1]

$\frac{8}{40} = 0.2$ [1]

units $= \frac{\text{units on } y\text{-axis}}{\text{units on } x\text{-axis}} = \frac{\text{cm}^3}{\text{min}}$

cm^3/min [1]

Answers

Q5. Δy = 10 g/dm^3 – 0.0 g/dm^3 = 10 g/dm^3

Δx = 240 s – 0 s = 240 s [1]

gradient = $\frac{\Delta y}{\Delta x}$

gradient = $\frac{10}{240}$ [1]

$\frac{10}{240}$ = 0.042 [1]

units = $\frac{\text{units on } y\text{-axis}}{\text{units on } x\text{-axis}}$ = $\frac{\text{g/dm}^3}{\text{s}}$

g/dm^3/s [1]

Mark scheme for chemistry questions

Q1. a) Δy = 6 – 1 = 5

Δx = 5 – 0 = 5 [1]

gradient = $\frac{\Delta y}{\Delta x}$

gradient = $\frac{5}{5}$ [1]

$\frac{5}{5}$ = 1 [1]

b) 1 [1]

Q2. Δy = 10 cm^3 – 0.0 cm^3 = 10 cm^3

Δx = 30 s – 0 s = 30 s [1]

gradient = $\frac{\Delta y}{\Delta x}$

gradient = $\frac{10}{30}$ [1]

$\frac{10}{30}$ = 0.33 (accept 0.3) [1]

units = $\frac{\text{units on } y\text{-axis}}{\text{units on } x\text{-axis}}$ = $\frac{\text{cm}^3}{\text{s}}$

cm^3/s [1]

Q3. a) Δy = 30 cm^3 – 0.0 cm^3 = 30 cm^3

Δx = 8 s – 0 s = 8 s [1]

gradient = $\frac{\Delta y}{\Delta x}$

gradient = $\frac{30}{8}$ [1]

$\frac{30}{8}$ = 3.75 (accept 3.8) [1]

units = $\frac{\text{units on } y\text{-axis}}{\text{units on } x\text{-axis}}$ = $\frac{\text{cm}^3}{\text{s}}$

cm^3/s [1]

b) Δy = 50 cm^3 – 36 cm^3 = 14 cm^3

Δx = 20 s – 10 s = 10 s [1]

gradient = $\frac{\Delta y}{\Delta x}$

gradient = $\frac{14}{10}$ [1]

$\frac{14}{10}$ = 1.4 [1]

units = $\frac{\text{units on } y\text{-axis}}{\text{units on } x\text{-axis}}$ = $\frac{\text{cm}^3}{\text{s}}$

cm^3/s [1]

Q4. Δy = 0.016 mol – 0.011 mol = 0.005 mol

Δx = 80 s – 25 s = 55 s [1]

gradient = $\frac{\Delta y}{\Delta x}$

gradient = $\frac{0.005}{55}$ [1]

$\frac{0.005}{55}$ = 0.00009 or 9 × 10^{-5}

(accept 0.000091 or 9.1 × 10^{-5}) [1]

units = $\frac{\text{units on } y\text{-axis}}{\text{units on } x\text{-axis}}$ = $\frac{\text{cm}^3}{\text{s}}$

mol/s [1]

Q5. a) Δy = 32 cm^3 – 24 cm^3 = 8 cm^3

Δx = 80 – 20 = 60 °C [1]

gradient = $\frac{\Delta y}{\Delta x}$

gradient = $\frac{8}{60}$ [1]

$\frac{8}{60}$ = 0.13 [1]

cm^3/°C [1]

b) 21.3 cm^3 [1]

Mark scheme for physics questions

Q1. a) Δy = 4 – 1 = 3

Δx = 1 – 0 = 1 [1]

gradient = $\frac{\Delta y}{\Delta x}$

gradient = $\frac{3}{1}$ [1]

$\frac{3}{1}$ = 3 [1]

b) 1 [1]

Q2. Δy = 8.00 m – 0.00 m = 8.00 m

Δx = 10 s – 0 s = 10 s [1]

gradient = $\frac{\Delta y}{\Delta x}$

gradient = $\frac{8.00}{10}$ [1]

$\frac{8.00}{10}$ = 0.8 [1]

units = $\frac{\text{units on } y\text{-axis}}{\text{units on } x\text{-axis}}$ = $\frac{\text{m}}{\text{s}}$

m/s [1]

Q3. a) Δy = 3 m/s – 0 m/s = 3 m/s

Δx = 2 s – 0 s = 2 s [1]

gradient = $\frac{\Delta y}{\Delta x}$

gradient = $\frac{3}{2}$ [1]

$\frac{3}{2}$ = 1.5 [1]

units = $\frac{\text{units on } y\text{-axis}}{\text{units on } x\text{-axis}} = \frac{\text{m/s}}{\text{s}} = \frac{\text{m}}{\text{s} \times \text{s}}$

m/s² [1]

b) Δy = 5 m/s – 3 m/s = 2 m/s

Δx = 6 s – 2 s = 4 s [1]

gradient = $\frac{\Delta y}{\Delta x}$

gradient = $\frac{2}{4}$ [1]

$\frac{2}{4}$ = 0.5 [1]

units = $\frac{\text{units on } y\text{-axis}}{\text{units on } x\text{-axis}} = \frac{\text{m/s}}{\text{s}} = \frac{\text{m}}{\text{s} \times \text{s}}$

m/s² [1]

Q4. Δy = 0 m/s – 6 m/s = –6 m/s

Δx = 6 s – 0 s = 6 s [1]

gradient = $\frac{\Delta y}{\Delta x}$

gradient = $\frac{-6}{6}$ [1]

$\frac{-6}{6}$ = –1 [1]

units = $\frac{\text{units on } y\text{-axis}}{\text{units on } x\text{-axis}} = \frac{\text{m/s}}{\text{s}} = \frac{\text{m}}{\text{s} \times \text{s}}$

m/s² [1]

Q5. a) Δy = 12 m/s – 2 m/s = 10 m/s

Δx = 40 s – 20 s = 20 s [1]

gradient = $\frac{\Delta y}{\Delta x}$

gradient = $\frac{10}{20}$ [1]

$\frac{10}{20}$ = 0.5 [1]

units = $\frac{\text{units on } y\text{-axis}}{\text{units on } x\text{-axis}} = \frac{\text{m/s}}{\text{s}} = \frac{\text{m}}{\text{s} \times \text{s}}$

m/s² [1]

b) Δy = 0 m/s – 12 m/s = –12 m/s

Δx 50 s – 40 s = 10 s [1]

gradient = $\frac{\Delta y}{\Delta x}$

gradient = $\frac{-12}{10}$ [1]

$\frac{-12}{10}$ = –1.2 [1]

units = $\frac{\text{units on } y\text{-axis}}{\text{units on } x\text{-axis}} = \frac{\text{m/s}}{\text{s}} = \frac{\text{m}}{\text{s} \times \text{s}}$

m/s² [1]

Tangents

Mark scheme for faded examples

Q1. Calculate answer:

$\frac{26}{60}$ = 0.43 (rounded from 0.43333333) [1]

Write units: cm³/s [1]

Q2. Use tangent to find values for Δy and Δx:

Δy = 11.2 cm³ – 6.0 cm³ = 5.2 cm³

Δx = 40 s – 10 s = 30 s [1]

Substitute values into equation:

gradient = $\frac{\Delta y}{\Delta x}$

gradient = $\frac{5.2}{30}$ [1]

Calculate answer:

$\frac{5.2}{30}$ = 0.17 (accept 0.173) [1]

Write units:

units = $\frac{\text{units on } y\text{-axis}}{\text{units on } x\text{-axis}} = \frac{\text{cm}^3}{\text{s}}$

cm³/s [1]

Mark scheme for chemistry questions

Q1. Δy = 12 cm³ – 4 cm³ = 8 cm³

Δx = 40 s – 0 s = 40 s [1]

gradient = $\frac{8}{40}$ [1]

$\frac{8}{40}$ = 0.2 [1]

cm³/s [1]

Q2. Δy = 1.12 g – 0.6 g = 0.52 g

Δx = 100 s – 25 s = 75 s [1]

$\frac{0.52}{75}$ [1]

$\frac{0.52}{75}$ = 0.0069 (accept 0.006 93) [1]

g/s [1]

Q3. Δy = 60 cm³ – 22 cm³ = 38 cm³

Δx = 24 s – 0 s = 24 s [1]

gradient = $\frac{38}{24}$ [1]

$\frac{38}{24}$ = 1.58 (accept 1.6 or 1.583) [1]

cm³/s [1]

Q4. Δy = 0.017 mol – 0.005 mol = 0.012 mol

Δx = 50 s – 0 s = 50 s [1]

gradient = $\frac{0.012}{50}$ [1]

$\frac{0.012}{50}$ = 0.00024 [1]

mol/s [1]

Q5. Suitable tangent drawn on graph at 50 seconds

e.g. from (0,24) to (100,56) [1]

Tangent used to find values for Δy and Δx e.g.

$\Delta y = 56$ cm³ – 24 cm³ = 32 cm³

$\Delta x = 100$ s – 0 s = 100 s [1]

Correct substitution e.g.

gradient = $\frac{32}{100}$ [1]

Correct calculation e.g.

$\frac{32}{100} = 0.32$ [1]

cm³/s [1]

Q6. Suitable tangent drawn on graph at 40 seconds

e.g. from (0,20) to (76,60) [1]

Tangent used to find values for Δy and Δx

e.g. $\Delta y = 60$ cm³ – 20 cm³ = 40 cm³

$\Delta x = 76$ s – 0 s = 76 s [1]

Correct substitution e.g.

gradient = $\frac{40}{76}$ [1]

Correct calculation e.g.

$\frac{40}{76} = 0.53$ (accept 0.526) [1]

cm³/s [1]

Area under a velocity-time graph

Mark scheme for faded examples

Q1. Find the area of the shapes

area = $\frac{1}{2}$ × 5 × 15

area = 37.5 [1]

Add units:

37.5 m [1]

Q2. Find the area of the shapes:

Distance travelled during first 30 seconds:

area = $\frac{1}{2}$ × base × height

area = $\frac{1}{2}$ × 30 × 20

area = 300

Distance travelled during the last 30 seconds:

area = length × width

area = 30 × 20

area = 600 [1]

Find the total area:

total area = 300 + 600 = 900 [1]

Add units:

900 m [1]

Q3. Find the area of the shapes:

area of one square: area = 10 × 2

area = 20

Estimate the total number of squares:

approximately 24 squares (accept 23–25) [1]

Find the total area:

area = number of squares × area of one square

area = 24 × 20

area = 480 [1]

Add units:

480 m [1]

Q4. Find the area of the shapes:

area of one square:

area = length × width = 5 × 2

area = 10

Estimate the total number of squares:

approximately 20 squares [1]

Find the total area:

area = number of squares × area of one square

area = 20 × 10 = 200 [1]

Add units: 200 m [1]

Mark scheme for physics questions

Q1. area = 5 × 10 = 50 m [2]

Q2. area = $\frac{1}{2}$ × 6 × 6 = 18 m [2]

Q3. a) area = $\frac{1}{2}$ × 3 × 25 = 37.5 m [2]

b) area = 3 × 25 = 75 m [2]

Q4. area = 2 × 10 = 20 m

area = $\frac{1}{2}$ × 5 × 10 = 25 m

total area = 20 + 25 = 45 m [3]

Q5. area = $\frac{1}{2}$ × 5 × 50 = 125 m

area = 2 × 50 = 100 m

area = $\frac{1}{2}$ × 1 × 50 = 25 m

total area = 125 + 100 + 25 = 250 m [3]

Q6. area = $\frac{1}{2}$ × 10 × 12 = 60 m [2]

Q7. area of one square = 10 × 5 = 50 m

approximately 12 squares

area = 12 × 50 = 600 m [3]

Q8. time = 3.5 × 60 = 210 s

area = 210 × 15 = 3150 m [3]

Geometry and trigonometry

Using angles

Mark scheme for faded example

Q1. Measure the length of the line and convert to N: 31.6 N

(accept an answer between 31 and 32 N) [1]

For the direction, measure the angle: 18° above horizontal (accept and answer between 18° and 19°) [1]

Mark scheme for physics questions

Q1. a) Angle labelled between the incident ray and normal. [1]

b) Angle labelled between refracted ray and normal. [1]

c) 30° (accept between 29° and 31°) [1]

d) 19° (accept between 18° and 20°) [1]

Q2. Horizontal **and** vertical forces drawn to the same scale. [1]

Resultant force drawn in the correct direction. [1]

magnitude = 100 N [1]

direction = 53° above horizontal [1]

Q3. Horizontal **and** vertical forces drawn to the same scale. [1]

Resultant force drawn in the correct direction. [1]

magnitude = 541 N [1]

direction = 56° above horizontal [1]

Q4. Horizontal **and** vertical forces drawn to the same scale. [1]

Resultant force drawn in the correct direction. [1]

magnitude = 1471 N [1]

direction = 55° above horizontal [1]

Area and surface area to volume ratio

Mark scheme for faded examples

Q1. area of a triangle $= \frac{1}{2} \times 63 = 31.5$ [1]

units = m^2 [1]

Q2. Substitute variables into equation:

area of rectangle = 20 × 15 = 300 [1]

units: cm^2 [1]

Mark scheme for biology questions

Q1. area of a rectangle = 25 × 25 = 625 [1]

units = cm^2 [1]

Q2. a) area of one face = 4.0 × 4.0 = 16.0 [1]

surface area of cube = 6 × 16.0 = 96.0 [1]

units = μm^2 [1]

b) volume of cube = 4.0 × 4.0 × 4.0 [1]

volume of cube = 64.0 [1]

units = μm^3 [1]

c) surface area : volume = 96 : 64 = 3 : 2 [1]

Q3. a) area of one face = 1.5 × 1.5 = 2.25 [1]

surface area of cube = 6 × 2.25 = 13.5 [1]

units = cm^2 [1]

b) volume of cube = 1.5 × 1.5 × 1.5 [1]

volume of cube = 3.375 [1]

units = cm^3 [1]

c) surface area : volume

= 13.5 : 3.375 = 4 : 1 [1]

d) area of one face = 3.0 × 3.0 = 9.0 [1]

surface area of cube = 6 × 9.0 = 54.0 [1]

units = cm^2 [1]

e) volume of cube = 3.0 × 3.0 × 3.0 [1]

volume of cube = 27.0 [1]

units = cm^3 [1]

f) surface area : volume = 54 : 27 = 2 : 1 [1]

Q4. Divide the rectangle into two. The largest rectangle has an area of:

45 × 30 = 1350 [1]

The smaller rectangle has an area of

20 × 15 = 300 [1]

Add the two areas: 1350 + 300 = 1650 [1]

units = m^2 [1]

Mark scheme for chemistry questions

Q1. a) area of one face = = 50 × 50 = 2500 [1]

surface area of cube = 6 × 2500 = 15 000 [1]

units = mm^2 [1]

b) volume of cube = 50 × 50 × 50 [1]

volume of cube = 125 000 [1]

units = mm^3 [1]

c) 3 : 25 [1]

Q2. a) area of one face = 2 × 2 = 4 [1]

surface area of cube = 6 × 4 = 24 [1]

units = nm^2 [1]

Answers

b) volume of cube = 2 × 2 × 2 [1]

volume of cube = 8 [1]

units = nm^3 [1]

c) 3 : 1 [1]

Q3. a) area of one face = 1.5 × 1.5 = 2.25 [1]

surface area of cube = 6 × 2.25 = 13.5 [1]

units = nm^2 [1]

b) volume of cube = 1.5 × 1.5 × 1.5 [1]

volume of cube = 3.375 [1]

units = nm^3 [1]

c) 4 : 1 [1]

Mark scheme for physics questions

Q1. a) area of one face = 5 × 5 = 25 [1]

surface area of cube = 6 × 25 = 150 [1]

units = cm^2 [1]

b) volume of cube = 5 × 5 × 5 [1]

volume of cube = 125 [1]

units = cm^3 [1]

Q2. area of one surface = 50 × 50 = 2500 [1]

units = mm^2 [1]

Q3. volume of cube = 25 × 25 × 25 [1]

volume of cube = 15 625 [1]

units = mm^3 [1]

Q4. area of faces = 2(2 × 2) + 4(2 × 3) [1]

surface area of cube = 32 [1]

units = cm^2 [1]

Q5. area of a triangle = $\frac{1}{2}$ × 8.4 × 6.5 [1]

= $\frac{1}{2}$ × 54.6 = 27.3 [1]

units = m^2 [1]